Student's Guide to

Masterton and Slowinski's

CHEMICAL PRINCIPLES

SECOND EDITION

RAY BOYINGTON

Department of Chemistry,
University of Connecticut

WILLIAM L. MASTERTON

Professor of Chemistry,
University of Connecticut

 SAUNDERS GOLDEN SERIES

1975   W. B. SAUNDERS COMPANY / Philadelphia / London / Toronto

W. B. Saunders Company: West Washington Square
Philadelphia, PA 19105

12 Dyott Street
London, WC1A 1DB

833 Oxford Street
Toronto, Ontario M8Z 5T9, Canada

The front cover illustration is a black and white reproduction of *Variation within a Sphere, No 10 The Sun*, a metal sculpture, by Richard Lippold, Courtesy The Metropolitan Museum of Art, Fletcher Fund, 1956.

Student's Guide to Masterton and Slowinski's Chemical Principles   ISBN 0–7216–1901–0

Last digit is the print number: 9   8   7   6   5   4   3   2   1

Preface

This study guide has been written as an aid to the student of general chemistry — to supplement a textbook and lecture series or to guide independent study. The chapter sequence and selection of topics parallel that of *Chemical Principles,* 3d edition, by W. L. Masterton and E. J. Slowinski.

Each chapter begins with a set of questions. These are meant to suggest some of the things you should be looking for in studying the topics of this chapter and the corresponding text and lecture material. Next follows a list of concepts and mathematical manipulations with which you should be familiar in order to get the most out of this unit of study. A chapter summary then presents an overview of some of the ideas treated in this unit.

A major change in this edition of the Study Guide is the addition to each chapter of a section entitled "Basic Skills." Here are listed the simple concepts that must be mastered before you can work problems in the text or on an examination. Each skill is illustrated by an example, either directly or by reference to an appropriate example and corresponding problems in the text. At the end of the section you will find a set of simple problems numbered to correspond directly to each skill. Hopefully, you should be able to work these with little difficulty. Following them (set off by a horizontal line) are several more complex problems, where you must decide for yourself which skill to apply. In several chapters, we have included a few "bonus problems" (set off by a second horizontal line), which we trust you will find intriguing. Answers to all problems are given in the appendix at the back of the study guide.

A feature of the first edition which students found most useful is the "self-test" included with each chapter, which consists of true-false and multiple choice questions. We have added a few questions to these tests (and changed a few answers!). Finally, a brief annotated list of carefully selected and recommended readings is given — for more basic, background study; for interesting and important applications; and for in-depth study.

The major objectives, then, in writing this guide have been to provide you with problems and tests by which you can measure your understanding of the material; to help you apply what you learn to new problems and to suggest further avenues of study; and to help you acquire a feel for the chemist's view of the physical world, an acquaintance with its limitations and its potentials.

SUGGESTIONS FOR THE USE OF THIS GUIDE

First, let's agree that what you get out of this course of study depends in great part on what you are willing to put into it. What follow are a few suggestions as to how you can make the most of your own efforts.

There are certain basic skills that you will frequently need to rely on, possibly further develop. The most fundamental of these are reading with comprehension and solving mathematical problems once they have been set up. There is probably no better way to learn or improve a language skill than through repeated usage, and that is one reason for this guide. Also, the problems and the tests in this guide will help you to decide just what you do and do not understand. As for the math, there are likely to be very few calculations in this course which you cannot do. These may consist of working with exponential notation; working with logarithms; solving first order (linear) equations; or solving less frequently encountered second order (quadratic) equations. For these you may simply need some review. Suggestions for background readings follow; additional help is generally yours for the asking, from fellow students and from your instructor.

1. Attend lectures and take notes, at least at first. Notes are important in answering questions such as these: What seem to be the important points? What is the lecturer's emphasis? What new examples and applications are given? What kind of experiments are done or cited? *Think* about what is being said, do not merely take notes.

2. Prepare for the lecture! Read over the material you expect to be discussed. Try to get a general idea of what the material is about. (Some lecturers assume a basic preparation, some do not. Even if the lecturer tells you nothing new, you will have learned more by having done the work yourself.) The more you know about what the lecturer is saying, the more you will be able to learn; the fewer notes you will need to take, and the better they will be; the more questions you can raise; or, the more time you can devote to other studies.

3. Get as much as you can out of discussions with other students and with your instructor, out of discussion or problem-solving classes — by being prepared, and by understanding (first by discovering for yourself) the inherent limitations and potentials of your class. (You will generally find, for example, that an instructor is better able to explain specific points and

answer specific questions than he is to guess what problems you are having.) Make your problems known, as early and as clearly as possible. Take advantage of the skills and insights of your instructor and your fellow students! You usually do have a very real influence on how beneficial a discussion class may be. Remember that this guide is designed to help you evaluate your progress for yourself, to recognize those areas where you need help, and to assist you in finding it; but, it is no substitute for the classroom.

4. Review regularly and frequently, if only to avoid cramming for an exam. This is where your notes should be most helpful and hopefully this guide as well. Keep this use in mind when taking notes from lecture or text. Look for connections between topics.

5. Read more chemistry — whether out of curiosity or from a feeling of helplessness. The more you do, the more rewarding this study must be.

To sum up:

Before attending lecture: Skim through the textbook chapter, reading introductory paragraphs and section headings, looking at tables and graphs; read and *add your own notes* to the first three sections of the study guide chapter. Then go back and *work* through the textbook chapter. In this and all other parts of the course, you've got to *actively participate.* Try to anticipate answers; write in questions you need answered; jot down your own conclusions in your own words.

Before discussing problems: Work through as many problems as possible in this guide and in your text; work the self-test in the guide; note anything requiring further explanation; think about your answers — do they seem reasonable in terms of what you already know? Do they suggest other problems?

Before taking exams: Review the main concepts and their physical meaning by referring to your notes and to the chapter summaries; recall and try to anticipate the emphases. Work some more problems.

For further assistance or enjoyment: Ask questions of your fellow students and your instructor. Do some reading. Teach what you have learned.

ADDITIONAL STUDY AIDS AND SUGGESTED READINGS*

Math Preparation and Problem-Solving Manuals

Butler, I. S., and A. E. Grosser, *Relevant Problems for Chemical Principles,* Menlo Park, Calif., W. A. Benjamin, 1970.
   Lots of interesting problems with detailed solutions and commentary; often applied to realistic situations.

---

*Besides magazine articles, readings suggested are generally available in paperback.

Masterton, W. L., and E. J. Slowinski, *Elementary Mathematical Preparation for General Chemistry,* Philadelphia, W. B. Saunders, 1974.
   Chemical problems are given throughout, with solutions; extensive discussion of math; of continuing usefulness.

Peters, E. I., *Problem Solving for Chemistry,* Philadelphia, W. B. Saunders, 1971.
   Worked problems are "programmed" to encourage your participation; answers given for somewhat basic problems.

Pierce, C., and R. N. Smith, *General Chemistry Workbook,* San Francisco, W. H. Freeman, 1971.
   Solutions given to about half of the problems; includes discussion of some math.

Risen, W. M., Jr., and G. P. Flynn, *Problems for General and Environmental Chemistry,* New York, Appleton-Century-Crofts, 1972.
   Lots of problems of varying difficulty; solutions given to all.

Sienko, M. J., *Chemistry Problems,* 2nd ed., Menlo Park, Calif., W. A. Benjamin, 1972.
   Lots of worked examples; problems with answers; particularly well-written, straightforward; some math discussed.

Programmed Instruction

Barrow, G. M., and others, *Understanding Chemistry,* New York, W. A. Benjamin, 1969.
   Self-teaching, self-testing; good background and practice material for most of the topics covered here.

Journals and Magazines

Chemical and Engineering News
   C & EN is a weekly newsmagazine of the profession, published by the American Chemical Society; much is generally readable, from letters and editorials to special reports.

Chemistry
   A monthly variety of readings of diverse levels and quality; might be worth a subscription. (Chemistry, 1155 16th St., N. W., Washington, DC 20036; $7/yr or group rate.)

Journal of Chemical Education
   J Chem Ed is another professional journal; monthly; of some interest to beginning students, particularly for resource papers and latest approaches to teaching.

Scientific American
   Occasional articles of chemical or applied interest in this monthly magazine; generally of top quality.

Audio-Visual Aids

Tapes, films, slides, programmed instruction materials and other resources that might be used for self-instruction are often available but not used. Find out where they are and how to use them, and whether they are worth your time. Often available at any library. Reprints of published papers are often available free from the author; sometimes, for a small fee, from the publisher.

Handbooks

Two very useful reference books, primarily compilations of chemical and physical properties and frequently updated, are:
Handbook of Chemistry, *N. Lange, Ed., New York, McGraw-Hill.*
Handbook of Chemistry and Physics, *R. C. Weast, Ed., Cleveland, Chemical Rubber.*

This book is dedicated to you, the student of general chemistry. We welcome your criticisms and your suggestions for the many ways there must be for making this more useful.

Contents

Basic Concepts

QUESTIONS TO GUIDE YOUR STUDY

1. What role does chemistry play in today's world? How might the role change in the near future? (Can you recall any chemical issues recently debated in the news media?)

2. What kinds of problems do chemists try to solve? Are there any specially useful or simplifying approaches to their solution? Is there a chemists' "point of view"?

3. What kinds of materials does the chemist generally work with in the laboratory? (Are they, for example, the materials of the "real" world, like steel, plastic, nylon, beer, and tobacco smoke?)

4. How does the chemist obtain, prepare, and purify the materials he studies?

5. What are some of the instruments and techniques of the chemist?

6. What properties or changes in properties does the chemist measure and use to describe materials? How are these measurements communicated to other scientists and to the public?

7. What limitations and uncertainties are there in these measurements? How are these communicated?

8. What kind of test would you perform in your kitchen to show that a sample of baking soda is "pure"? (What is the chemical meaning of "pure"?)

9. What are some of the unsolved problems — the frontiers — in chemistry?

10. What questions would you want to ask of a government-employed chemist?

11.

12.

YOU WILL NEED TO KNOW*

Concepts

Though no previous encounter with chemistry is assumed at this point, at least a general understanding is assumed for the notions of matter, energy, composition, experiment, measurement . . . .

Math

1. How to use exponential ("scientific") notation — Appendix 4.
2. How to recognize and solve first order (linear) equations, like $y = ax + b$. — Readings.

CHAPTER SUMMARY

You know, perhaps all too well, that we live in an age of science and technology, an age that is now two to three centuries old yet very much with us. For example, nearly ninety percent of all scientists that there have ever been are alive now. At least through the first half of this century science seemed, by any measure, to be growing exponentially: about every ten to fifteen years the number of scientific journals approximately doubled and the number of compounds known to chemists likewise doubled. (Try estimating the number of papers in chemistry expected in the year 2000, knowing that there were about 50,000 in 1950.)

This course of study will introduce you to the underlying principles and methods of one of the more fruitful areas of scientific endeavor, an endeavor involving many thousands of persons in many nations.

But just what does a chemist do? Most of us would probably agree that a chemist is one who is qualified to determine the feasibility of chemical change. A major objective of the chemist is to be able to predict the conditions under which a chemical reaction may occur and to describe the course the reacting system may take. (Does this suggest the kind of social role the chemist might play? After more than ten years of research, and applying a well known chemical principle, F. Haber was able to describe the conditions under which atmospheric nitrogen could be converted to

---

*All chapters and appendices referred to are in *Chemical Principles*; otherwise, see the readings list for appropriate background material.

ammonia. A process made commercial in 1913, the Haber preparation accounts for most of the ammonia produced, whether for eventual use in fertilizers, explosives, or Everybody's Ammonia Cleanser.)

The history of modern chemistry really began with the introduction of quantitative experimentation, with measurement. The idea of composition, barely mentioned here, is taken up in more detail in the next two chapters. There are several levels of meaning to *composition*: one speaks of atomic composition, for example; or elemental composition, or composition by weight, and so forth. All of these usages refer to the way in which component parts or building blocks, whether simple in themselves or very complex, are put together to form matter as we know it.

A major objective for the student in an introductory course is to see how the principles of chemistry are experimentally established and then applied in predictions of reaction feasibility. The chemical systems that we will look at will be simple in composition (there will generally be only one or two components) and simple in the number of things that happen (only one physical or chemical change will be considered).

BASIC SKILLS

In this introductory chapter you are expected to become familiar with some of the common types of measurements that chemists make and the experiments they carry out to separate and identify substances. You should, for example, be familiar with the measurement of mass and the units in which masses of substances are commonly expressed. Again, you should understand the principles that govern such separation techniques as distillation and chromatography. In another area, you should be able to explain how such properties as boiling point and absorption spectrum can be used to identify a substance and check its purity.

A few specific skills are required in this chapter. They include the following.

1. Apply the rules of significant figures to calculations based upon experimental measurements.

These rules are discussed on pp. 6–8, and illustrated both in the body of the text and in Examples 1.1 and 1.2. Students ordinarily have little trouble learning these rules, but they often forget to apply them in carrying out chemical calculations. None of the problems at the end of Chapter 1 refer directly to significant figures, but in each case you are expected to report an answer to the correct number of significant figures. This will be true throughout the text and this study guide.

2. Use the conversion factor approach to convert lengths, volumes, or masses from one unit to another.

This is a general approach which can be applied to a wide variety of problems in chemistry. It will be used extensively in later chapters of the text. If you are not familiar with it, this chapter gives you an excellent opportunity to become skilled in its use.

The basic principle behind the conversion factor approach is discussed on p. 9, and is illustrated in some detail in Example 1.3. Notice, in part (b) of this example, how it can be applied to conversions involving more than one step. The example that follows further illustrates this principle.

- -

The density of mercury is 13.6 g/cm^3. What is its density in kg/m^3? ___

Clearly, two conversions are required: mass must be changed from g to kg and volume from cm^3 to m^3. The first conversion factor is available directly in Table 1.1.

$$(1) \quad 1 \text{ kg} = 10^3 \text{ g}$$

The second factor does not appear directly in the table. However, we can readily obtain it from the relation: 1 m = 10^2 cm. Cubing both sides of this equality:

$$(2) \quad (1 \text{ m})^3 = (10^2 \text{ cm})^3 \,; 1 \text{ m}^3 = 10^6 \text{ cm}^3$$

Now we are ready to set up the problem:

$$13.6 \; \frac{\text{g}}{\text{cm}^3} \times \frac{1 \text{ kg}}{10^3 \text{ g}} \times \frac{10^6 \text{ cm}^3}{1 \text{ m}^3} = 13.6 \times 10^3 \; \frac{\text{kg}}{\text{m}^3} = 1.36 \times 10^4 \; \frac{\text{kg}}{\text{m}^3}$$

$$\qquad\qquad\quad (1) \qquad\quad (2)$$

Notice that the conversion factors are set up in such a way (1 kg/10^3g; 10^6 cm^3/1 m^3) that the original units (g/cm^3) cancel out, leaving the desired units (kg/m^3). You will also note that here, as in many other calculations in chemistry, you must be familiar with the rules of exponential notation (Appendix 4).

- -

Many of the problems at the end of Chapter 1 require conversions of one type or another. Perhaps the most straightforward of these are Problems 1.9 and 1.22. You should also be able to work Problems 1.10, 1.11, 1.23, and 1.24 without much difficulty using the conversion factor approach.

3. Relate the mass and volume of a sample of matter to its density.

"Density problems" can be worked directly from the defining equation:

$$\text{density} = \text{mass/volume}$$

See Example 1.2 and the following example.

What is the volume of a sample of aluminum (d = 2.70 g/ml) weighing 12.0 g? _____

Rearranging the defining equation to solve for volume:

$$\text{volume} = \text{mass/density}$$

Substituting the quantities given in the statement of the problem:

$$\text{volume} = (12.0 \text{ g})/(2.70 \text{ g/ml}) = 4.44 \text{ ml}$$

Several of the problems at the end of Chapter 1 require that you be familiar with the density relation (1.19, 1.20, 1.21).

4. Relate temperatures expressed on one scale to those on another scale.

The three temperature scales commonly used in general chemistry ($^\circ$F, $^\circ$C, and $^\circ$K) are related by the equations:

$$^\circ\text{F} = 1.8\,^\circ\text{C} + 32^\circ$$

$$^\circ\text{K} = \,^\circ\text{C} + 273^\circ$$

These equations illustrate the fact that there is always a linear relationship between temperature scales, i.e.,

$$^\circ y = a^\circ x + b$$

where a and b are constants. When y is plotted vs. x, a is the slope (i.e., $\Delta y/\Delta x = a$) and b is the "y intercept" (i.e., the value of y when $x = 0$). From Figure 1.4, we see that in the equation relating $^\circ$F to $^\circ$C:

$$a = \frac{212 - 32}{100 - 0} = \frac{180}{100} = 1.8; b = 32$$

Similarly, for the relationship between $^\circ$K and $^\circ$C, $a = 1$ and $b = 273$.

Most of the temperature conversions that you will be required to make in general chemistry involve °C and °K; occasionally °F may be involved. Such conversions can be carried out very simply with the two equations cited above. Problems 1.12 and 1.25 are somewhat more subtle in that they require that you apply the general linear relationship between temperature scales. A logical approach is indicated in the following example.

_ _

Suppose a scale of temperatures is set up so that 50°C = 100°Y and 150°C = 500°Y. What is the general equation relating °Y to °C?

One way to work this problem would be to construct a graph of °Y vs. °C by plotting the two points given and drawing a straight line through them. The slope of that straight line would give the value of the constant a in the equation:

$$°Y = a°C + b$$

while the intercept b could be determined by extrapolating to find the value of °Y when °C = 0.

The two constants can be obtained with less effort by applying simple algebra to the basic equation. Using subscripts 2 and 1 to represent the higher and lower temperatures respectively:

$$°Y_2 = a°C_2 + b; 500° = 150°a + b$$

$$°Y_1 = a°C_1 + b; 100° = 50°a + b$$

To solve for a, we subtract the second equation from the first.

$$400° = 100°a; \ a = 4.00$$

To obtain b, we substitute for a in one of the equations, e.g.,

$$500° = (4.00)150° + b; \ b = 500° - 600° = -100°$$

The general equation must then be: $°Y = 4.00°C - 100°$

_ _

5. Use a table of solubilities (such as Table 1.2) to design an experiment to separate two solids by fractional crystallization.

The general approach is described in the discussion on pp. 13–14, and is applied in Problems 1.13 and 1.26. In working these problems note that:

— solubilities at a given temperature can be treated like conversion factors. Thus, to work 1.13a, we might consider that, at 80°C: 98 g tartaric acid ≏ 100 g water.

the solubility of one solid is assumed to be independent of the presence of the other. Thus, at 100°C, it is assumed that 100 g of water will dissolve 98 g of tartaric acid, regardless of how much succinic acid is present, and 71 g of succinic acid, regardless of how much tartaric acid is present. Obviously, this is an approximation, but it is probably more nearly accurate than any other simple assumption we might make at this point.

Problems

1. Assuming that each of the following involves inexact numbers, express answers to the correct number of significant figures.

   a) 12.62/1.28   b) 1.96 × 3.4112
   c) 2.9 × 12.41 / 8.62   d) 12.0 + 3.92 + 0.1

2. The density of bromine is 3.12 g/cm^3. Express its density in

   a) kg/cm^3 b) g/liter   c) kg/m^3 d) lb/in^3 e) lb/ft^3

3. Using the density given in (2), calculate

   a) the mass of 522 cm^3 of bromine.
   b) the volume of bromine which has a mass of one kilogram.

4. On a certain temperature scale, Y, the normal boiling point of water is 200°Y and the freezing point is 20°Y.

   a) Obtain a general relation between °Y and °C.
   b) Express 50°C in °Y; in °F; in °K.

*5. Use Table 1.2 to answer the following.

   a) How much water is required to dissolve 62 grams of tartaric acid at 80°C?
   b) How much succinic acid can be dissolved in 75 grams of water at 20°C?
   c) If the solution in a) is cooled to 40°C, how much tartaric acid will crystallize out of solution?

— — — — — — — — — — — —

6. A student determines the density of a piece of metal by weighing it (mass = 15.68 g) and adding it to a flask which has a volume of 25.00 cm^3. He finds that 21.20 g of water (d = 0.9970 g/cm^3) is required to fill the flask with the metal in it. Express, to the correct number of significant figures:

   a) the volume occupied by the water.
   b) the volume occupied by the metal.
   c) the density of the metal in g/cm^3.
   d) the density of the metal in lb/in^3.
   e) the volume, in cubic inches, of a sample of metal weighing 11.2 lb.
   f) the volume, in liters, of a sample of metal weighing 112 g.

7. Complete the following table.

°C	0	50	100	150	200	250
°X	-100	0	100	___	___	500
°F	32	___	___	___	400	___
°K	273	282	291	300	309	318

   (First use the two points given to obtain a relation between °C and °X)

- - - - - - - - - - - - - - -

8. The planet Azote has a sea of liquid ammonia, a mountain range of solid ammonia, and a thin atmosphere of gaseous ammonia. Naturally, the exotic inhabitants of Azote base their temperature scale on the melting point (0°N) and boiling point (100°N) of ammonia. An earthling's normal body temperature (98.6°F) would correspond to what reading, in degrees N, on the Azote scale? (Consult the text or a handbook of chemistry for the physical properties of ammonia.)

9. Suppose you wish to prepare a water solution containing 60 g of A and 20 g of B. The solubility of A, in g/100 g H_2O, depends on the Celsius temperature, t, according to the equation:

$$S(A) = 20 + 4.0 \times 10^{-1} t + 2.0 \times 10^{-3} t^2;$$

while that of B is given by:

$$S(B) = 10 + 3.0 \times 10^{-1} t + 4.0 \times 10^{-3} t^2.$$

a) What minimum amount of water should you use to dissolve the sample at 100°C?

b) To what temperature can the solution be lowered before *B* starts to crystallize out?

c) How much solid *A* appears at this temperature?

SELF-TEST

True or False

1. Changes in physical state, like melting and boiling, tend to resolve matter into pure component substances. ( )

2. Extensive properties characterize, chemically identify, a substance. ( )

3. In trying to identify a certain liquid compound "L", a student finds that its density, freezing and boiling points, absorption spectrum, and behavior in a chromatography column are indistinguishable from those of known compound "Z". The student can safely assume L and Z are one and the same compound. ( )

4. Nearly all countries have already adopted or are now adopting a metric system as the single recognized system of measurement. ( )

5. The simplification in using the metric system is that, when converting units within the system, a decimal point is moved. ( )

6. Appropriate conversion factors would allow you to convert from a volume measurement in cubic feet to a density measurement in grams per cubic centimeter. ( )

7. Since silicon is the second most abundant element in the earth's crust, it must be one of the cheapest to buy from a chemicals supplier. ( )

8. Sodium chloride and potassium dichromate, both solids which are soluble in water, can probably be separated by taking advantage of differences in solubility. ( )

9. A liter is almost equivalent to a quart. ( )

Multiple Choice

10. One way to definitely show that gasoline is a mixture of (   )
substances would be to
    (a) measure its density
    (b) measure the temperature during boiling
    (c) burn it
    (d) filter it

11. A procedure appropriate to the separation of the components (   )
of gasoline is:
    (a) burning   (b) fractional crystallization
    (c) fractional distillation   (d) column chromatography

12. The fact that behavior on melting may allow you to decide (   )
whether or not a sample of matter is a pure substance depends on the
general observation that
    (a) melting points can be measured with high accuracy
    (b) all pure substances can be melted
    (c) the melting temperature is sensitive to small amounts of
        impurities
    (d) the melting point does not vary with sample size

13. A sample of matter that exhibits uniform behavior during all (   )
changes in physical state would have to be
    (a) an element   (b) a compound
    (c) a mixture   (d) a pure substance

14. Which one is an example of an extensive property? (   )
    (a) volume   (b) color   (c) temperature   (d) density

15. A suitable, nondestructive test for determining whether or (   )
not a particular beautiful, green gem is an emerald would be:
    (a) a melting point determination   (b) see if it is scratched
    (c) measure its absorption spectrum (d) weigh it

16. Identification of the amino acid components in Super (   )
Crunchy Corn Flakes would probably involve
    (a) distillation   (b) recrystallization
    (c) filtration   (d) chromatography

17. For a given substance, density ordinarily increases in the (   )
order:
    (a) solid, liquid, gas   (b) liquid, solid, gas
    (c) gas, liquid, solid   (d) solid, gas, liquid

18. If radioactive dating reveals the age of a manuscript to be ( )
between 1150 and 1390 years, its age may be represented as
    (a) 1.270×10^3 yr   (b) 1.27×10^3 yr
    (c) 1.3×10^3 yr   (d) 1×10^3 yr

19. How many significant figures are there in the number $6.50 \times$ ( )
10^3?
    (a) two   (b) three   (c) four   (d) six

20. A steel screw weighs 0.50 ounce. Its mass in grams ( )
$(454 \text{ g} = 1 \text{ lb} = 16 \text{ oz})$ is:
    (a) $0.50 \times 454 \times 16$ (b) $0.50 \times 454/16$
    (c) $0.50 \times 16/454$ (d) none of these

21. If the volume and mass measurements on a sample of arsenic ( )
are 2.10 ml and 12.040 g, the reported value for the density of
arsenic, in g/ml, should have how many significant figures?
    (a) two   (b) three   (c) four   (d) five or more

22. How many times larger is a Kelvin degree than a Fahrenheit ( )
degree?
    (a) 1.8   (b) 1/1.8   (c) 32   (d) they are the same

23. It is found that the chirping frequency of a tree cricket, f, is ( )
approximately related to the Celsius temperature by the equation:
$°C = 0.1\ f + 8$. One would expect that the chirping frequency and the
Fahrenheit temperature would be
    (a) unrelated
    (b) equal to each other
    (c) related by the equation for a straight line
    (d) directly proportional to each other

SELF-TEST ANSWERS

    1. T   (The basis for fractional distillation, for example.)
    2. F   (Intensive properties do.)
    3. T   (Identity of properties must mean identity of composition.)
    4. T
    5. T
    6. F   (Volume and density are different kinds of measurement.)
    7. F   (It is not very readily isolated from its compounds.)
    8. T   (The basis for fractional crystallization.)
    9. T
   10. b   (A constant temperature would be characteristic of a pure
         substance, as a general rule.)

11. c   (The bulk of the mixture consists of liquid substances.)
12. c
13. d   (Either a compound or an element would behave this way.)
14. a
15. c
16. d   (See Chapter 23.)
17. c   (Water is exceptional; Chapter 11.)
18. c   (The second digit is uncertain by one unit; the third, even more.)
19. b   (This notation shows that the 6.50 contains these digits.)
20. b
21. b
22. a
23. c   (Try substituting $0.1 f + 8$ for $°C$ in: $°F = 1.8 °C + 32$.)

SELECTED READINGS

Bachmann, H. G., The Origin of Ores, *Scientific American* (June 1960), pp. 146–156.
  A discussion of the abundance and availability of elements; an introduction to geochemistry.
Edman, D. D., and others, Computers and Chemistry, *Chemistry* (January 1972), pp. 6–9.
  Beginning a series of articles on computers and programming of chemical problems.
Keller, R. A., Gas Chromatography, *Scientific American* (Oct. 1961), pp. 58–67.
  A closer look at a very important analytical and preparative tool.
Masterton, W. L. and E. J. Slowinski, *Elementary Mathematical Preparation for General Chemistry*, Philadelphia, W. B. Saunders, 1974.
  A good place to go for math background, practice exercises.
Price, D. J. de S., *Little Science, Big Science*, New York, Columbia University Press, 1963.
  On the exponential growth of science and its possible implications.
Ritchie-Calder, Lord, Conversion to the Metric System, *Scientific American* (July 1970), pp. 3–11.
  Light reading on the history and adoption of the metric system.
Storms, H. A., and W. Tichy, Probing Concentration Zero, *Chemistry* (March 1973), pp. 6–10.
  A graphic discussion of very sensitive analytical tools.
Ziman, J. M., *Public Knowledge*, Cambridge, Cambridge U. Press, 1968.
  Discusses the interactions between the scientist and his colleagues; or, how the scientific community arrives at a consensus and calls it scientific knowledge. Readable though not easy.

2

Atoms, Molecules and Ions

QUESTIONS TO GUIDE YOUR STUDY

1. Can you think of common, everyday observations which suggest that matter is made of atoms and molecules; or that it isn't?

2. Why do neutral atoms exist? (Why not a world made up of ions?) Or, why do ions form, some bearing positive charge, others negative?

3. How would you experimentally show that sulfur and oxygen combine in a one-to-one weight ratio to form the gaseous compound sulfur dioxide? How would you demonstrate the conservation of matter?

4. Just how small are atoms and molecules? Is there a convenient way of counting them and of expressing their numbers; of weighing them?

5. If more than one kind of atom exists for a particular element (e.g., strontium-90 is but one kind of strontium atom), then how are we to interpret the numbers representing the weights of atoms?

6. What holds atoms together in a molecule? Could you explain it to a first grader?

7. What are the supporting arguments, based on observation, for the atomic theory of matter?

8. How do you distinguish, in terms of atomic theory, between an element and a compound; or, between one iron chloride and another?

9. If the idea of atoms is at least as old as ancient Greece, then why did it take so long for the atomic theorists to get anywhere?

10. If not an atom, then what is the fundamental, ultimate building-block? (Does the question even make sense?)

11.

12.

YOU WILL NEED TO KNOW

See this section of Chapter 1, Study Guide.

CHAPTER SUMMARY

With more than a little help from his friend the physicist, the chemist has solved the problem of the nature of bulk matter and the changes it undergoes. He interprets ("explains") the unique properties of substances in terms of the unique properties of invisibly small particles that constitute matter: atoms, ions and molecules. The directly observable behavior of matter is the collective behavior of very large numbers of these particles. This imaginative view of matter provides the framework for most of the explanation, and success, of modern science. The biologist extends this theory to deal with very large molecules; the geologist, to account for the large-scale processes occurring within the earth; the astronomer, to explain the birth and death of stars.

It is a long way to go from an individual molecule of water to a clean mountain brook, or even a beaker of distilled water; but all the fundamental principles appear to be known. These principles comprise "atomic theory" and "kinetic molecular theory," and were established between 1800 and 1930, approximately. During that time, and since then, the theory has remained dynamic — growing in its detail and sharpness of focus with every new experiment. Atomic theory is the unifying theme of this course.

Now, the chemist certainly doesn't think of atoms as "ultimate" particles; but atoms, and to a certain extent, part of their underlying structure, are as fundamental, as far down the ladder, as he need consider. (The high energy reactions of modern physics, well outside the range of reactions considered to be chemical, have turned up two hundred subatomic particles, or thereabouts. Except for electrons and nuclei, our only encounter with such interesting and extraordinary matter will be to mention neutrons — and then only to explain the fact that atomic weights must be considered as average values, and to find that low energy, "chemical" neutrons sometimes bring about interesting reactions. More about this latter topic in Chapter 22.)

The arrangements of electrons and nuclei in atoms, ions, and molecules form the theoretical basis of all chemistry. Their properties seem sufficient for explaining bulk samples of matter. All observable changes are thought of as being changes in the arrangements of electrons and nuclei within atoms, ions and molecules; or as rearrangements of the atoms, ions and molecules themselves; or both. (This elaboration of atomic theory begins mainly with Chapter 6 and continues throughout the text.)

The laws of chemical combination (all of the important ones, like the law of constant composition, were known over a century ago) provide some rather indirect support for these ideas. More compelling support for atomic theory, including the existence of particles smaller than atoms, comes from relatively recent experiments: electrical discharge through gases; atomic and molecular spectra (Chapter 6); x-ray diffraction by crystalline solids (Chapter 9); and radioactivity (Chapter 22).

How do you measure out a number of material objects, say three dozen apples? Everyday measurements would include counting, as well as weighing and determining volume. A small number of apples you no doubt would count off. But what if you needed three thousand? You would probably weigh them out. Conceivably, you could even measure their volume. Well, these same three kinds of measurement are the only ways in which the chemist can normally deal with bulk matter consisting of vast numbers of particles. One of the most important things for him to know, then, is the relationship between mass and volume of a sample and the number of particles the sample contains. Volume measurements are typical in the case of handling gases (Chapter 5) or liquids and solutions (Chapter 10); weighing is the only other everyday measure. Even the largest molecules are too small to be directly counted: there is no chemical counterpart to everyday counting. Hence the need for relating masses and numbers of particles.

The gram atomic weight of an element, in particular of isotope ^{12}C, is a convenient reference sample for measuring mass **and** determining the number of particles present in the sample, all at the same time. Several independent methods of determining the number of atoms in exactly 12 grams of ^{12}C give a value of 6.02×10^{23}. Avogadro's number is to the chemist what the ream of paper is to the typist, and the dozen to the poultry farmer. It is the number of atoms found in one GAW of *any* element; the number of molecules in one gram molecular weight of *any* molecular substance. It is the number of structural or formula units known to have a certain mass, a gram formula weight.

Another use for this relationship is seen in the way the chemist finds it convenient to represent reactions: the chemical equation, as you will see in the next chapter, treats the reaction of bulk matter as though individual atoms, molecules, and ions were taking part, step by step, to give the overall, net or observable, result. Equations show how we imagine counting off particles, and this we have done by measuring a mass.

BASIC SKILLS

1. Relate the number of protons and neutrons in an atom to its nuclear symbol.

The nuclear symbol gives the *atomic number* as a subscript at the lower left of the symbol of the element and the *mass number* as a superscript at the upper left.

atomic number = number of protons

mass number = number of protons + number of neutrons

This skill is shown in Example 2.1 and is required to work Problems 2.2 and 2.19. In working these problems, note that in a neutral atom the number of electrons is equal to the number of protons; a positive ion is formed by the loss of electrons, a negative ion by a gain of electrons.

2. Given the atomic weights of two elements, calculate the relative (average) masses of their atoms.

- -

What is the ratio of the mass of a strontium atom (AW = 87.62) to that of a beryllium atom (AW = 9.012)? _____
The masses of atoms are in the same ratio as their atomic weights. Hence, a Sr atom must be 87.62/9.012 = 9.723 times as heavy as a Be atom.

- -

This skill is required to work Problems 2.7 and 2.24.

3. Given the masses and abundances of the isotopes of an element, calculate its average atomic weight. Perform the reverse calculation.

The first type of calculation is illustrated in Example 2.2; Problem 2.9 is entirely analogous. The same skill is involved in a somewhat more subtle way in Problem 2.26, where you are required to obtain the ratio of the average atomic weight of oxygen to that of its most abundant isotope, ^{16}O.
The calculation of isotopic abundances from the average atomic weight of an element, given the masses of the isotopes, is illustrated in the following example.

- -

Thallium (AW = 204.37) consists of two isotopes, ^{203}Tl (mass = 203.05), and ^{205}Tl (mass= 205.05). What is the % abundance of the heavier isotope? _____

Let us represent the fraction of the heavy isotope, ^{205}Tl , by x. Since the fractions of the two isotopes must add up to one, that of the light isotope must be $(1 - x)$. The equation for the atomic weight becomes:

$$AW\ Tl\ = 204.37 = x(205.05) + (1 - x)\ (203.05)$$

Solving: $204.37 = 205.05x + 203.05 - 203.05x = 2.00x + 203.05$

$$2.00x = 204.37 - 203.05 = 1.32; \quad x = 132/2.00 = 0.660$$

Converting to percent, we find the abundance of ^{205}Tl to be 66.0%.

– –

Problem 2.10 is entirely analogous to the example just worked. Problem 2.27 is a little more difficult in that three isotopes are involved. Note, however, that you are given the abundance of one of these isotopes.

4. Relate numbers of gram atomic weights and gram molecular weights to masses in grams and to numbers of identical particles.

Conversions of this type are used extensively throughout the general chemistry course. They are illustrated in Examples 2.3 and 2.4. Note that:

— the conversion factor relating gram atomic weight to number of atoms or gram molecular weight to number of molecules is the same for all substances.

$$1\ GAW\ H = 6.023 \times 10^{23}\ H\ atoms$$

$$1\ GAW\ Mg = 6.023 \times 10^{23}\ Mg\ atoms$$

$$1\ GMW\ H_2O = 6.023 \times 10^{23}\ H_2O\ molecules$$

— the conversion factor relating gram atomic weight or gram molecular weight to mass in grams depends upon the substance involved.

$$1\ GAW\ H = 1.008\ g$$

$$1\ GAW\ Mg = 24.31\ g$$

$$1\ GMW\ H_2O = 2(1.008\ g) + 16.00\ g = 18.02\ g$$

Problems 2.11 and 2.28 give you practice in working with conversions of this type. Problems 2.12 and 2.29 are somewhat more subtle but involve the same principle. Note further that the same skill, along with other conversions, is required in Problems 2.13 to 2.15 and 2.30 to 2.32. In Problem 2.15, for example, a logical first step would be to calculate the number of Si atoms in one mg.

*5. Use weight analysis data for two or more compounds of two elements to illustrate the Law of Multiple Proportions.

_ _

The percentages by weight of the elements in two different oxides of copper are:

compound (1): % Cu = 88.8, % O = 11.2

compound (2): % Cu = 79.9, % O = 20.1

What are the weights of copper per gram of oxygen in the two compounds? _____ , _____

How do these numbers support the Law of Multiple Proportions? _____

In 100 g of compound (1), we have 88.8 g of Cu and 11.2 g of O. The weight of copper per gram of oxygen must be:

88.8 g Cu/11.2 g O = 7.93 g Cu/g O

Similarly, in compound (2) we have:

79.9 g Cu/20.1 g O = 3.98 g Cu/g O

The two numbers obtained are in the ratio 2:1 (i.e., 7.93/3.98 = 1.99). In other words, twice as much copper is combined with a given mass of oxygen in compound (1). The Law of Multiple Proportions predicts that some such simple ratio should be obtained; in another case, it might be 3:1, 3:2, etc.

_ _

Problem 2.5 can be worked in this same manner; Problem 2.22 is a variation on the same theme.

*6. Given the voltages required to bring two ions of the same charge to the same point in the mass spectrometer, calculate the ratio of their masses.

The basis for this type of calculation is Equation 2.2, which in a more general form might be written:

$$m_x E_x = m_y E_y$$

where x and y represent two different nuclei. On p. 37, this relation is used to obtain the atomic weight of He; in Problem 2.8, it is applied to find the mass of a phosphorus isotope. Problem 2.25 requires that you calculate the ratio E_x/E_y, given m_x and m_y.

*7. Relate the specific heat of a metal to its approximate gram atomic weight.

The basic relation here is Equation 2.3, a form of the Law of Dulong and Petit.

- -

The specific heat of calcium is 0.155 cal/g°C. Estimate its GAW. _____
Substituting in Equation 2.3:

$$GAW \approx \frac{6 \text{ cal/}°C}{\text{specific heat}} \approx \frac{6 \text{ cal/}°C}{0.155 \text{ cal/g}°C} \approx 40 \text{ g}$$

- -

The Law of Dulong and Petit is, of course, only approximate (AW Ca = 40.1). Problem 2.16 shows how one can use the law, along with analytical data for a compound of the metal in question, to obtain a quite accurate value for the atomic weight.

Problems

1. a) Give the number of neutrons, protons and electrons in a $^{60}_{27}$Co atom; the +2 ion derived from $^{90}_{38}$Sr.
   b) Write symbols for nuclei containing 12 p + 11 n; 48 neutrons, atomic number 37; 82 neutrons, mass number 139. (Use the Periodic Table to find symbols of elements.)
2. The average mass of an atom of element X is 7.12 times as great as the mass of an atom of ^{12}C. What is the atomic weight of X?
3. a) Silicon consists of three isotopes of masses 27.99, 28.99, and 29.98; the abundances of these isotopes are 92.28, 4.67 and 3.05% respectively. Calculate an average atomic weight for silicon.

b) Chlorine consists of two isotopes of masses 34.98 and 36.98 respectively. Taking the average atomic weight of chlorine to be 35.45, calculate the % abundances of the two isotopes.

4. Complete the following table.

	AW or MW	no. grams	no. GAW or GMW	no. atoms or molecules
Mg	——	——	——	2.1×10^3
HCl	——	——	1.64	——
C_6H_6	——	52.0	——	——

*5. The element chromium forms three different oxides in which the percentages of chromium are 52.0, 68.4, and 76.5 respectively. Calculate the number of grams of chromium per gram of oxygen in each compound and show how the numbers you obtain illustrate the Law of Multiple Proportions.

*6. In measuring the atomic weight of a certain isotope, it is found that the voltage required to bring a +1 ion to the collector is 502 volts; with a +1 ion derived from ^{12}C, the corresponding voltage is 795 volts. Calculate the atomic weight of the isotope.

*7. Using Equation 2.3, estimate the specific heat of manganese ($AW = 55$).

— — — — — — — — — — — —

8. Neon consists of three isotopes of masses 20.00, 21.00, and 22.00; its average atomic weight is 20.18. The abundance of ^{21}Ne is 0.28%. What are the abundances of the other two isotopes?

9. If the atomic weight scale were set up taking the average atomic weight of hydrogen to be exactly 1, what would be:
   a) the average atomic weight of carbon?
   b) the atomic weight of ^{12}C?
   c) the molecular weight of H_2O?

10. How many atoms are there in
   a) 26.0 ml of Hg ($d = 13.6$ g/ml; AW = 200.6)?
   b) 4.06 grams of benzene, which has the molecular formula C_6H_6?
   c) 5.63 GAW of ytterbium?

11. What is the mass of grams of
   a) a molecule of nitric acid (molecular formula HNO_3)?
   b) 1.50×10^{-10} gram atomic weights of Cl?
   c) a solution containing one gram molecular weight of CH_3OH in 24.0 ml of H_2O ($dH_2O = 1.00$ g/ml)?

SELF-TEST

True or False

1. The chemical properties of an atom are determined by its   (   )
nuclear charge.

2. In the light of current knowledge, we must view the law of   (   )
constant composition as only a good approximation to observed
behavior.

3. An isotope is one of two or more atomic species having the   (   )
same atomic number but different numbers of electrons.

4. The following statement is probably consistent with modern   (   )
atomic theory: a positively charged sodium ion is smaller than a
neutral sodium atom.

5. The molecular weight of a substance is simply a number   (   )
which tells how heavy a molecule is when compared to a chosen
reference.

6. A gram molecular weight of hydrogen (two atoms in a   (   )
molecule) is a gram of hydrogen.

7. The number of molecules in a gram molecular weight of   (   )
water, 18 grams, is eighteen times the number of atoms in a gram
atomic weight of hydrogen, 1 gram.

8. All neutral atoms of a given element have the same number   (   )
of electrons.

9. Stable, bulk samples of matter carry little or no electric   (   )
charge; that is, they are electrically neutral.

10. The atomic weight of chlorine is 35.5. Chlorine consists of   (   )
two isotopes of mass numbers 35 and 37. It follows that the Cl-35
isotope must be the more abundant.

11. The atomic weight is an average number that takes into   (   )
account all known isotopes of an element, including those prepared
artificially in the laboratory.

Multiple Choice

12. Which of the following has the smallest mass?   (   )

    (a) an atom of C   (b) a molecule of CO_2
    (c) one gram of C   (d) one GAW of C

13. A certain isotope X has an atomic number of 7 and a mass ( )
number of 15. Hence,

(a) X is an isotope of   (b) X has 8 neutrons per
     nitrogen   atom
(c) an atom of X has 7   (d) all the above
     electrons

14. The relative abundances of ^{35}Cl and ^{37}Cl in a sample of ( )
elementary chlorine can be determined by

(a) fractional crystallization
(b) vapor phase chromatography
(c) precipitation with silver nitrate
(d) mass spectrometry

15. To illustrate the Law of Multiple Proportions, we could use ( )
data giving the percentages of

(a) magnesium in magnesium oxide and in magnesium
     chloride
(b) carbon in carbon monoxide and in carbon dioxide
(c) potassium dichromate in two different mixtures with
     sodium chloride
(d) ^{63}Cu and ^{65}Cu isotopes in copper metal

16. Suppose the atomic weight scale had been set up with ( )
calcium, Ca, chosen for a mass of exactly 10 units, rather than about
40 on the present scale. On such a scale, the atomic weight of oxygen
would be about

(a) 64   (b) 32   (c) 16   (d) 4

17. The most direct method for determining atomic weights is: ( )

(a) gas density measurements
(b) mass spectrometry
(c) combining weights
(d) α-particle scattering

18. An element is: ( )

(a) a collection of atoms with identical numbers of neutrons
(b) a collection of atoms with identical nuclear masses
(c) a collection of atoms with identical nuclear charges
(d) one of the following: air, earth, fire, water

19. Experimental support for the existence of atoms and sub-   (   )
atomic particles includes all of the following except:

(a) radioactive decay
(b) electrical discharge through gases
(c) metal foil scattering of alpha particles
(d) continuous mechanical subdivision of a single crystal

20. By about how many orders of magnitude (powers of ten) is   (   )
the diameter of an atom larger than that of a nucleus?

(a) 1   (b) 2   (c) 4   (d) 10,000

21. There is always a ratio of small whole numbers between: (   )

(a) the gram atomic and gram molecular weights of an
    element
(b) the weight percentage of copper in any two of its
    compounds
(c) the weights of copper combined with one gram of
    element A in CuA and one gram of element B in CuB.
(d) the atomic weights of any two elements

22. The diameter of an atom of gold is of the order of   (   )

(a) 10^{-5} cm   (b) 10^{-8} cm   (c) 10^{-12} cm   (d) 10^{-23} cm

23. The mass of an individual atom is of the order of   (   )

(a) 10^{-22} kg   (b) 1/2000 g   (c) 10^{-22} g   (d) 6×10^{23} g

24. The idea that most of the mass of an atom is concentrated in   (   )
a very small core, the nucleus, is a result of the experiments of

(a) Dalton   (b) Bohr   (c) Cannizzaro   (d) Rutherford

25. The charge on the nucleus of a neon atom is: (   )

(a) +20   (b) +10   (c) zero   (d) −10

26. The reason that the densities of two different gases at the   (   )
same temperature and pressure compare as their respective molecular
weights is:

(a) all gases have the same molecular weight
(b) all gases have the same density at the same temperature
    and pressure
(c) all gas molecules are monatomic
(d) assuming Avogadro to be right, each gas density is
    directly proportional to the mass of the respective gas
    molecule

27. Sufficient information for your calculation of the simplest   (   )
atom ratio of nitrogen to hydrogen in the compound hydrazine
would be:
    (a) the atomic weights of nitrogen and hydrogen
    (b) the ratio of masses for nitrogen and hydrogen atoms and
        the composition by weight of hydrazine
    (c) the combining volumes of nitrogen and hydrogen in
        forming hydrazine
    (d) the atomic weights of nitrogen and hydrogen, and the
        weight composition of several other nitrogen-hydrogen
        compounds

28. In one gram molecular weight of hydrogen, H_2, there are   (   )
Avogadro's number of:
    (a) hydrogen atoms   (b) electrons
    (c) hydrogen molecules   (d) neutrons

SELF-TEST ANSWERS

  1. T
  2. T   (We do find some atom ratios are not ratios of small whole
        numbers and that some ratios are at least slightly variable.)
  3. F   (Different numbers of neutrons.)
  4. T   (As you might guess, removing one or more electrons subtracts
        from the total volume of the atom, which is mostly electrons
        anyway.)
  5. T
  6. F   (2.0 grams.)
  7. F   (There are 6.02×10^{23} molecules in one GMW, 6.02×10^{23}
        atoms in one GAW.)
  8. T
  9. T   (This observation is often called the principle of electro-
        neutrality.)
10. T
11. F   (Only naturally occurring isotopes.)
12. a
13. d
14. d
15. b
16. d   (The weights of the individual atoms still compare 40/16, or 10/4.)
17. b
18. c
19. d

20. c   (Atomic, 10^{-8} cm; nuclear, 10^{-12} cm.)
21. a   (Molecules contain whole numbers of atoms.)
22. b
23. c   (Divide any GAW by N.)
24. d
25. b   (Same as atomic number.)
26. d   (For a given volume, you are comparing equal numbers of molecules.)
27. b   (More about formulas in Chapter 3.)
28. c

SELECTED READINGS

Causey, R. L., Avogadro's Hypothesis and the Duhemian Pitfall, *J. Chem. Ed.* (June 1971), pp. 365-367.
  How and why Avogadro's idea came not to be accepted for some fifty years, even by the best of his colleagues.
Feinberg, G., Ordinary Matter, *Scientific American* (May 1967), pp. 126-134.
  A survey of the history of ideas about the structure of matter.
Lagowski, J. J., *The Structure of Atoms,* Boston, Houghton Mifflin, 1964.
  Useful here and for Chapter 6; presents excerpts and comments on experimental support for atomic theory.
Lavoisier, A., *Elements of Chemistry,* New York, Dover, 1965.
  A principles approach to chemistry as known in 1789; used by Dalton in his teaching. It contains the first(?) statement of the law of conservation of matter.
Lucretius, *On the Nature of the Universe,* Baltimore, Penguin, 1951.
  A two thousand year old classic that addresses you in the framework of a highly speculative atomic theory.
Patterson, E. C., *John Dalton and the Atomic Theory,* Garden City, N. Y., Doubleday-Anchor, 1970.
  A biography that is interesting for its re-creation of the social and scientific setting for Dalton's contributions.

Chemical Formulas and Equations

QUESTIONS TO GUIDE YOUR STUDY

1. What information is conveyed by a chemical formula; a chemical equation? Is there more than one kind of chemical formula; chemical equation?

2. What kind of experiment would you do to show that the formula of water is H_2O, and not HO as Dalton believed? What assumptions, if any, do you need to make?

3. Do you need to know the masses of individual atoms and molecules to be able to write chemical formulas and equations?

4. How does the chemist conveniently deal with numbers of reacting molecules; with their masses?

5. How do you represent the physical state (solid, liquid, solution . . .) of the materials taking part in a reaction?

6. Are the conditions, like temperature and pressure, under which a reaction occurs represented by the chemical equation for the reaction?

7. What does a chemical equation tell you about what you would see as a reaction proceeds? (For example: what shape, size and color crystal is formed in a certain reaction, and how fast?)

8. What does a chemical equation tell you about what the molecules are doing as a reaction proceeds?

9. What happens when you use nonstoichiometric amounts (quantities other than those represented by the chemical equation) in carrying out a reaction?

10. Can a reaction be of any use if you do not know the equation for the reaction? How would you experimentally establish the equation for a reaction?

11.

12.

YOU WILL NEED TO KNOW

Concepts

1. The meaning of gram atomic weight, gram molecular weight — Chapter 2

2. How to interpret formulas; in particular, the meaning and use of subscripts, parentheses and brackets. These ideas are implicit in Chapter 2 — see Chapter 3 Summary and Self-Test, Study Guide; Readings.

Example: The formula P_4O_{10} refers to a molecule in which there are *four atoms of phosphorus combined with ten atoms of oxygen.* As you will see in this chapter, the formula itself may refer to just one unit of structure, a molecule in this case, or it may refer to Avogadro's number of structural units — depending on the context in which the formula is used. But in any context, the atom ratio is four to ten in the structural unit.

The formula $Ca(NO_3)_2$ refers to a unit of structure in which there are *one Ca atom, two N atoms, and six O atoms;* or to Avogadro's number of such formula units. (The parentheses set off atoms which together may act as a unit; the subscript 2 then refers to two such groups of atoms.)

Math

1. How to work with conversion factors — Chapter 1
2. How to deal with significant figures — Chapter 1
3. How to relate gram formula weights (GAW, GMW), masses and numbers of particles — Chapter 2

CHAPTER SUMMARY

Stoichiometry is the awesome label usually attached to the arithmetic of chemistry. It is the quantitative interpretation and use of chemical formulas and equations. The practical importance of stoichiometry is easily illustrated — the principles established in this chapter will be applied throughout the text and in any quantitative laboratory work you do. And beyond this course: the mining engineer may want to know the amount of iron he can expect to extract from a given amount of the ore hematite, Fe_2O_3; the botanist may wish to determine the volume of oxygen liberated when a certain mass of glucose is photosynthesized; the weight-watcher may want to compare the energies stored in equal masses of carbohydrate and fat.

This chemical arithmetic is not particularly difficult or complicated. Its mastery may require considerable practice. First, it involves special labels and units: grams, formula weights, moles, and others. Second, the calculations will always be carried out using conversion factors instead of proportions (which many people would rather use, if only out of habit). This work-saving "mole method" of solving stoichiometric problems involves a consistent interpretation of all amounts of chemical species (atoms, ions, molecules, electrons, . . .) in terms of a single unit, the *mole.*

A mole measures a unit amount of a single chemical species or substance. This means two things: a mole is a *number* of particles or formula units; a mole represents a definite *mass,* associated with this number of particles.

(1) A mole is a counting unit with exactly the same kind of meaning and uses as other, more familiar counting units like *dozen.* A mole is Avogadro's number of identical items; the dozen, twelve items. Where we have used the words "gram atomic weight," "gram molecular weight," and "gram formula weight" we now substitute the one word "mole."

(2) A mole is thus a variable unit of mass, its value (in grams) depending on the formula weight.

> Consider that a "dozen identical objects" could just as well mean, simultaneously, a certain mass of objects and their number. If a penny weighs 3.0 grams, then counting out a dozen pennies is equivalent to weighing out 36 grams of pennies. Any amount of pennies, specified in terms of the unit dozen, will automatically mean a certain number of pennies and a certain mass of pennies. A unit amount (i.e., a dozen) of dimes will contain the same number of coins as a dozen pennies but will have a different mass (about 27 grams). We could conveniently count *or* weigh any amount of identical coins in units of dozen. In particular, we could be sure that, say, 72 grams of pennies and 54 grams of dimes contained equal numbers of coins.

The mole concept thus provides the chemist with the means of obtaining equal numbers of formula units by making the convenient measurement of mass. For two substances, he needs to weigh out masses that compare in the same way as do the formula weights. (Of course, the numbers of particles desired may not be equal: you would simply weigh out masses such that the *numbers* of gram formula weights, or moles, compare in the same way as do the desired numbers of particles or formula units. For example, satisfy yourself that 71 grams of chlorine, Cl_2, contain twice as many molecules as does 1.01 grams of hydrogen, H_2.)

The mole method of solving problems might be outlined as follows:

(1) Write the balanced chemical equation for the reaction.

(2) Use the coefficients of the equation as conversion factors for relating the numbers of moles of given and desired species.

(3) Convert the given amounts to moles by using the appropriate gram formula weights (i.e., grams per mole).

(4) Convert from moles of desired species to mass, if needed, by using gram formula weights, or to numbers of particles by using Avogadro's number (i.e., particles or formula units per mole), thus expressing the result in the units called for.

More often than not, the units of the desired quantity will indicate the kind of calculation you need perform. Suppose, for example, that you are given the mass of a reactant and are asked to find the mass of a certain product in a reaction. The series of conversions must take you from "grams reactant" through a relation of the numbers of moles of reactant and product, finally to the "grams product." The conversions called for in applying the mole method would look like this:

$$\text{g product} = \text{g reactant} \times \underbrace{\frac{1 \text{ mole reactant}}{\text{GFW reactant}}}_{\text{I}} \times \underbrace{\frac{\text{no. moles product}}{\text{no. moles reactant}}}_{\text{II}} \times \underbrace{\frac{\text{GFW product}}{1 \text{ mole product}}}_{\text{III}}.$$

Note that step I converts grams to moles; step II is the ratio of the coefficients gotten from the balanced equation; step III gives the desired units, converting from moles to grams.

BASIC SKILLS

1. Relate the number of moles of a species of known formula to the number of grams or the number of particles.

As pointed out in the discussion on p. 53, the term "mole" is always associated with a chemical formula. When the formula represents a single particle (e.g., Na, CO_2) the mole represents Avogadro's numbers of particles (6.02×10^{23} Na atoms, 6.02×10^{23} CO_2 molecules). The mass in grams of a mole can always be found by summing the gram atomic weights of the elements in the formula (1 mole Na = 23.0 g; 1 mole CO_2 = 12.0 g + 2(16.0 g) = 44.0 g).

Conversions between moles, grams and numbers of particles are illustrated in Example 3.5; Problems 3.10 and 3.28 are entirely analogous. Problems 3.9 and 3.27 further test your understanding of the mole concept. The fact that the mole can represent 6.02×10^{23} particles of any type (electrons, protons, water droplets, or potatoes) is emphasized in Problems 3.11 and 3.29.

2. Given the formula of a compound, calculate the percentages by weight of the elements.

_ _

What are the percentages by weight of hydrogen, nitrogen, and oxygen in nitric acid, HNO_3? _____

In one mole of HNO_3 there are

$$
\begin{aligned}
1\ \text{GAW H} &= \ \ 1.01\ \text{g} \\
1\ \text{GAW N} &= 14.00\ \text{g} \\
3\ \text{GAW O} &= \underline{48.00\ \text{g}} \\
&\ \ \ \ 63.01\ \text{g}
\end{aligned}
$$

The weight percentages must be:

$$
\text{H:}\frac{1.01}{63.01}\times 100 = 1.60\%;\ \text{N:}\frac{14.00}{63.01}\times 100 = 22.22\%;\ \text{O:}\frac{48.00}{63.01}\times 100 = 76.18\%
$$

_ _

A similar calculation is involved in Problems 3.1b and 3.19b. Also, Problems 3.4 and 3.22 are readily analyzed if you first obtain the percentages of Zn in ZnS, Cd in CdS, and C in aspirin.

3. Determine the simplest formula of a compound, given the percentages by weight of the elements or analytical data from which these percentages can be calculated.

A simple illustration of the determination of a simplest formula from percentage composition is given in Example 3.1. Example 3.2 shows how the percentages of the elements can be calculated from data obtained by chemical analysis. Finally, Example 3.3 shows a two-step calculation, first of percentage composition and then of a simplest formula.

It is worth pointing out that you can go directly from analytical data to simplest formula without passing through percentage composition. Consider, for instance, Example 3.3. Having decided that the original sample must contain 1.518 g of Cl, it follows that it must also contain (2.159 – 1.518) g = 0.641 g Sc. Working directly with these two quantities:

$$
\text{no. of GAW Sc} = 0.641\ \text{g Sc} \times \frac{1\ \text{GAW Sc}}{44.96\ \text{g Sc}} = 0.0143\ \text{GAW Sc}
$$

$$
\text{no. of GAW Cl} = 1.518\ \text{g Cl} \times \frac{1\ \text{GAW Cl}}{35.45\ \text{g Cl}} = 0.04282\ \text{GAW Cl}
$$

Clearly these numbers are in a 1:3 ratio, so the simplest formula must be $ScCl_3$.

The calculation of simplest formula from % composition is illustrated in Problems 3.2 and 3.20. In Problems 3.3 to 3.8 and 3.21 to 3.26, analytical data are given from which you should be able to obtain simplest formulas, either by working through percent composition, as in Example 3.3 of the text, or more directly, as explained above.

4. Write and balance simple equations.

This skill is described in Section 3.4 of the text (and again below), where the balanced equation is derived for the reaction of N_2H_4 with N_2O_4. A somewhat simpler example, dealing with a reaction in water solution, is presented here.

––

Write a balanced equation to represent the reaction that takes place when water solutions containing silver ions, Ag^+, and sulfide ions, S^{2-}, are mixed to give a precipitate of silver sulfide. _____

In order to write the balanced equation, we must first deduce the formula of the solid product. To achieve electrical neutrality, two Ag^+ ions are required to balance one S^{2-} ion. The simplest formula of silver sulfide must then be Ag_2S. The balanced equation follows directly:

$$2Ag^+(aq) + S^{2-}(aq) \rightarrow Ag_2S(s)$$

The symbol (aq) is used to indicate that the reactants, Ag^+ and S^{2-} ions, come from water solutions. These might, for example, be water solutions of silver nitrate, $AgNO_3$, and sodium sulfide, Na_2S.

––

Later in the course, we shall have a great deal more to say about writing and balancing equations. Equations of the type just written, often referred to as net ionic equations, are discussed in some detail in Chapter 16 of the text. A systematic method of balancing rather complex oxidation-reduction equations is described in Chapter 20. With a little bit of luck, you should be able to use the simple approach outlined in this chapter to work problems such as 3.12 and 3.30. At most, you will be expected to know the formulas and ordinary physical states of such common substances as carbon dioxide, water, and oxygen.

5. Given a balanced equation, relate the numbers of moles, grams, or particles of two substances taking part in the reaction.

The conversions required here are illustrated in some detail in Examples 3.6 and 3.7. Notice the consistent use of conversion factors here, as in previous chapters. Experience over many years convinces us that this approach is the most general and reliable way of analyzing a wide variety of problems in general chemistry. If you are still using the "ratio and proportion" method, it's high time you switched. Most of the problems that you will be assigned for homework or asked to solve on exams simply cannot be worked by a rote approach. Besides, it will help you to understand the examples worked here and in the text if we are using the same language.

Problems 3.13 to 3.16 and 3.31 to 3.34 involve mole-mole, mole-gram, or gram-gram conversions based on balanced equations. Often, one or more additional steps may be required. In Problem 3.32, for example, you might first calculate how many moles of H_2SO_4 there are in the solution and then apply the balanced equation to relate moles of H_2SO_4 to grams of FeS. Again, in Problem 3.15, you first have to find out how many grams of CO_2 are produced by three astronauts in six days. Problems 3.18 and 3.36 are somewhat more challenging but basically of the same type.

6. **Calculate the theoretical yield of product, given the numbers of grams or moles of two or more reactants.**

A suggested approach to this type of problem is illustrated in part (a) of Example 3.8. The key step here is to decide which substance is the limiting reagent upon which the theoretical yield must be based. The following example focuses on this step.

- -

In the reaction $2H_2(g) + O_2(g) \rightarrow 2H_2O(g)$, which elementary substance is the limiting reagent if 3.20 moles of H_2 are reacted with 2.40 moles of O_2? _____ if 1.00 gram of H_2 is reacted with 1.00 gram of O_2?_____

The balanced equation tells us that 2 moles of H_2 are required for every mole of O_2. Clearly, to react with 2.40 moles of O_2, we need twice as many, or 2(2.40) = 4.80 moles of H_2. Since we only have 3.20 moles of H_2, hydrogen must be the limiting reagent. From a slightly different point of view, we might reason that 3.20 moles of H_2 would require only half as many moles of O_2, i.e., 1.60 moles of O_2. Hence, oxygen must be in excess; 2.40 - 1.60 = 0.80 mole of O_2 will be left unreacted when the reaction is over.

To answer the second question asked above, it is convenient to calculate first the numbers of moles of H_2 and O_2.

$$1.00 \text{ g } H_2 \times \frac{1 \text{ mole } H_2}{2.02 \text{ g } H_2} = 0.495 \text{ mole } H_2$$

$$1.00 \text{ g } O_2 \times \frac{1 \text{ mole } O_2}{32.0 \text{ g } O_2} = 0.0313 \text{ mole } O_2$$

Clearly, we have a large excess of H_2 here; we need only 0.0626 mole H_2 to react with 0.0313 mole of O_2. Most of the 0.495 mole of H_2 will remain unreacted; the oxygen is the limiting reagent.

- -

This approach can be used, along with skill (5), to work Problems 3.17 and 3.35.

*7. Given the simplest formula of a substance and an approximate value for its molecular weight, obtain its molecular formula.

This skill is illustrated in Example 3.4. The principle involved is applied in the last part of Problems 3.5 and 3.23.

*8. Given the actual yield of product in a reaction and knowing (or having calculated) the theoretical yield, determine the percentage yield.

This type of calculation, illustrated in Example 3.8b, is readily carried out by using the definition:

$$\% \text{ yield} = \frac{\text{actual yield}}{\text{theoretical yield}} \times 100$$

See Problems 3.17c and 3.35b.

Problems

1. Complete the following table.

Formula	no. moles	no. grams	no. molecules	no. atoms
C	_____	_____		2.01×10^9
SiO_2	0.631	_____		
C_6H_6	_____	15.6	_____	
O_2	_____	_____	52.0	

2. Calculate the percentages by weight of the elements in

   a) CO   b) $AlCl_3$ c) $Al_2(SO_4)_3$

3. Determine the simplest formulas of

    a) a compound which contains 89.7% Bi and 10.3% O.
    b) a compound which contains 26.5% Cr, 24.5% S and 49.0% O.
    c) an organic compound containing only carbon and hydrogen, if combustion of a 1.000 g sample gives 3.030 g of CO_2.
    d) lead iodide, given that a sample weighing 0.752 g gives 0.766 g of AgI.

4. When an organic compound is burned in air, carbon is converted to CO_2 (g) and hydrogen to H_2O(l). Write balanced equations for the reactions that occur when the following compounds are burned.

    a) CH_4 (g)   b) C_6H_6 (l)   c) methyl alcohol, CH_3OH

5. The balanced equation for the reaction of N_2H_4 with N_2O_4 is given on p. 56 of the text. Determine:

    a) the number of moles of H_2O produced from 1.64 moles of N_2H_4.
    b) the number of grams of N_2H_4 required to form 9.16 moles of N_2.
    c) the number of grams of N_2O_4 required to react with 124 g of N_2H_4.
    d) the number of molecules of N_2 produced from 1.00 g of N_2O_4.

6. For the reaction $2Al(s) + 3Cl_2 (g) \rightarrow 2AlCl_3 (s)$, indicate which is the limiting reagent if one starts with:

    a) 1.00 mole of Al and 1.00 mole of Cl_2
    b) 12.0 g Al + 12.0 g Cl_2

*7. *In Problem 3c above, the molecular weight is approximately 60. What is the molecular formula of the compound?*

*8. *In Problem 6a above, what is the theoretical yield in grams of $AlCl_3$? If the actual yield is 60.0 g, what is the % yield?*

– – – – – – – – – – – –

9. An organic liquid contains only the three elements C, H, and O. Upon combustion of a 2.60 g sample, 3.81 g of CO_2 and 1.56 g of H_2O are formed.

    a) How many grams of carbon are there in the sample? How many grams of hydrogen? of oxygen?

    b) What are the percentages by weight of the elements in the organic liquid?

10. A sample of antimony chloride weighing 2.709 g is reacted with hydrogen sulfide to precipitate all of the antimony as Sb_2S_3; the precipitate weighs 2.018 g.

    a) What is the percentage of antimony in Sb_2S_3? (AW Sb = 121.8, S = 32.1)

    b) What is the weight of antimony in the sample?

    c) What are the percentages of antimony and chlorine in the sample?

    d) What is the simplest formula of antimony chloride?

11. Calculate the number of moles in

    a) 35.0 g of O_2 b) 1.60 x 10^{12} molecules of N_2

    c) 1.60 X 10^{12} Na^+ ions

12. 1.24 liters of a solution containing 0.532 mole/liter of Al^{3+} is mixed with 0.518 liter of a solution containing 3.16 moles/liter of OH^-. A precipitate of aluminum hydroxide forms.

    a) Write a balanced equation for the reactions in water solution.

    b) How many moles of Al^{3+} are present before reaction? how many moles of OH^-?

    c) What is the limiting reagent?

    d) What is the theoretical yield of aluminum hydroxide in moles? in grams?

– – – – – – – – – – –

13. A pure sample of manganese dioxide, MnO_2, weighing 0.435 g is heated. The only products are oxygen and another pure oxide of manganese, weighing 0.382 g.

    a) What is the simplest formula of the second oxide?

    b) Write a balanced equation for the reaction.

14. A certain solid is known to be either KCl or KI. When 1.00 g of the substance is dissolved and allowed to react with excess silver nitrate, 1.92 g of a precipitate (either AgCl or AgI) is formed. Determine by calculation whether the starting material was KCl or KI.

15. A 1.50 sample of $[Co(NH_3)_5 NO_2](NO_3)_2$, contaminated with some $NH_4 NO_2$, is fired in a stream of oxygen to yield 0.365 g of $Co_2 O_3$. Determine the purity of the original sample, assuming 100% conversion of the Co to $Co_2 O_3$.

NOTES ON WRITING CHEMICAL EQUATIONS

It may appear that you are being asked to write and understand chemical equations for lots of reactions when you don't yet know any chemistry. How, for example, do you know which substances are gases, which are solids? Or which substance is composed of molecules and which is not — and so what kind of formula to use? Or what conditions are required for the reaction to proceed as indicated by the equation? Our main concern at this point is with establishing a systematic approach to solving stoichiometric problems, and this does require at least some writing of balanced chemical equations. *The descriptive chemistry will be learned as you go along.*

So, what information do you need to be given and what do you need to find on your own, and where, so as to write equations? The following suggestions are to assist you in answering these and other questions. Further practice and suggestions can be found in looking ahead to Chapters 16, 18, 19 and 20, as well as to the Readings.

An equation represents, both qualitatively and quantitatively, observed changes in matter in terms of the rearrangements of atoms, ions, and molecules. An equation is "balanced" when it reflects the observed mass relationships between reactants and products, in particular the conservation of matter. Balancing requires that the same kinds and numbers of atoms appear in the products as started out in the reactants. To write a chemical equation ("balanced" is generally taken for granted):

1. You need to know, or be given, the reactants and products actually observed under the given reaction conditions.

At first, you will be given the names and formulas; gradually, you will acquire criteria for deciding what the substances are. Example: CO_2 and H_2O are the usual products in the "combustion" reactions of carbon- and hydrogen-containing compounds with oxygen or air.

You will need to use the molecular formula when a substance is known to be molecular; otherwise, the simplest formula. (Shouldn't the maximum information possible be incorporated in an equation?)

2. Show the physical states of all reactants and products.

These must be the states you would observe under the given reaction conditions. Example: If water is a product in a reaction occurring at room temperature, you would expect it to be a liquid.

3. Follow the various conventions such as writing the reactants (the substances consumed or disappearing during a reaction) to the left, products to the right, separated by an arrow ($\rightarrow$), and using the simplest ratio of whole numbers of formula units needed to balance the equation. (Balancing requires choosing the relative numbers of formula units so as to conserve atoms — not rewriting the formulas themselves! Example:

The formation of liquid water from hydrogen and oxygen gases, under almost all conditions, can be represented by the equation: $2 H_2 (g) + O_2 (g) \rightarrow 2 H_2 O(l)$ but *not* by the following "equation": $H_2 (g) + O(g) \rightarrow H_2 O(l)$, where the formula of oxygen has been manipulated so as to achieve a balance. (The formulas must correspond to reality: oxygen is known to be diatomic.)

Finally, be aware of some of the limitations of chemical formulas and equations. Beyond specifying the physical state, a formula in itself, or an equation, does not directly say anything about the conditions needed for the reaction to occur, or indeed whether the reaction can occur. An equation says nothing about reaction speed; nothing about the extent to which a given reactant is consumed; and, usually, nothing about how the reaction actually occurs among the individual atoms and molecules. All of these questions need to be answered by observations that are not represented by an equation.

SELF-TEST

True or False

1. Since three-fourths of the atoms in a sample of $ScCl_3$ are chlorine atoms, 75% of the weight of the sample is due to chlorine. ( )

2. The molecular formula for hydrogen peroxide, $H_2 O_2$, indicates that one molecule of hydrogen (H_2) is bonded to one molecule of oxygen (O_2). ( )

3. In 50 grams of calcium carbonate, $CaCO_3$ (FW = 100), there are: 3×10^{23} Ca atoms, 0.5 mole of C, and 3/2 gram atomic weights of oxygen. ( )

4. The calculation of theoretical yield is based on the assumption that all of the limiting reactant is consumed according to the equation for the reaction.   ( )

5. All of the following may be reasons why the actual product yield in a reaction is usually less than 100%: separation and purification results in losses; competing reactions form other products instead; the reaction hasn't stopped yet; not all the reactants are converted to desired product, even when the reaction has ceased.   ( )

6. The chemical mole is defined as being Avogadro's number of formula units (molecules, ions, atoms, electrons . . .).   ( )

7. In an ordinary chemical reaction, the number of moles of reactants always equals the number of moles of products.   ( )

Multiple Choice

8. The formula weight of lead iodide, PbI_2, is: ( )
   (a) $82 + 2(53)$ (b) $2(82 + 53)$
   (c) $207 + 2(127)$ (d) $207 + 127$

9. In analyzing a compound for carbon, the compound is burned in air and the masses of products determined. What assumption do we make?   ( )
   (a) all the oxygen in the air is converted to water, H_2O
   (b) all the carbon is converted to carbon dioxide, CO_2
   (c) equal numbers of moles of CO_2 and H_2O are produced
   (d) all the oxygen in the H_2O produced comes from the compound

10. An empirical or simplest formula of a substance always shows   ( )
   (a) the element(s) present and the simplest ratio of whole numbers of atoms
   (b) the actual numbers of atoms combined in a molecule of the substance
   (c) the number of molecules in a sample of the substance
   (d) the gram molecular weight of the substance

11. What *minimum* information would be sufficient for ( )
determining the simplest formula for a compound?
   (a) the elements present in the compound
   (b) the elements in the compound and their atomic weights
   (c) the elements in the compound, their atomic weights, and
       their combining weights in a sample of the compound
   (d) the elements in the compound, their atomic weights,
       their combining weights and weight percentages in a
       sample of the compound

12. In order to determine the composition of a substance, one ( )
might
   (a) determine its physical properties and compare them to
       those of known substances
   (b) convert the substance to one or more substances of
       known composition
   (c) compare the products for each of several reactions to
       those gotten when similarly reacting other, known
       substances
   (d) all of the above

13. Information given by the formula for hydrazine, N_2H_4, ( )
includes all of the following except:
   (a) hydrazine could just as well be represented by the
       formula NH_2
   (b) the per cent by weight that is nitrogen is (28.0/32.0) ×
       100%
   (c) one molecule of hydrazine contains six atoms
   (d) one mole of hydrazine weighs 32.0 grams

14. The formula for calcium carbonate ($CaCO_3$, FW = 100) ( )
generally represents all of the following except:
   (a) one formula unit ("molecule") of calcium carbonate
   (b) N formula units of calcium carbonate
   (c) one gram of calcium carbonate
   (d) one hundred grams of calcium carbonate

15. A compound with the simplest formula C_2H_5O has a molecu- ( )
lar weight of 90. The molecular formula for the compound is:
   (a) $C_3H_6O_3$ (b) $C_4H_{26}O$ (c) $C_4H_{10}O_2$ (d) $C_5H_{14}O$

16. Which contains the largest number of molecules? ( )
   (a) 1.0 g CH_4 (MW = 16)   (b) 1.0 g H_2O (MW = 18)
   (c) 1.0 g HNO_3 (MW = 63)   (d) 1.0 g N_2O_4 (MW = 92)

17. About how much oxygen is there in exactly one mole of   ( )
baking soda, $NaHCO_3$?
    (a) 16 g   (b) 24 g   (c) 48 g   (d) 96 g

18. It is estimated that there are 1×10^{21} kg of water in all the   ( )
oceans. How many moles of water is this?

    (a) $\dfrac{1 \times 10^{21}}{18}$ (b) $\dfrac{1 \times 10^{21} \times 10^3}{18}$

    (c) $1 \times 10^{21} \times 18 \times 10^3$ (d) $\dfrac{1 \times 10^{21}}{6 \times 10^{23}}$

19. The molecular weight of a protein that causes food poisoning   ( )
is about 900,000. The approximate mass of one molecule of this
protein is:
    (a) 2×10^{-18} g   (b) 1×10^{-6} g
    (c) 9×10^5 g   (d) some other number

20. A balanced chemical equation shows   ( )
    (a) the mole ratio in which substances react
    (b) the direction a chemical system will move in and the
        extent or yield of the reaction
    (c) the speed with which the reaction proceeds
    (d) the individual molecular steps by which the reaction
        occurs

21. To write a chemical equation for a given reaction, you would   ( )
need to know at least:
    (a) the masses of the reactants and products in the given
        reaction
    (b) the mole ratios of all reactants and products in the
        reaction
    (c) the formulas of all reactants and products in the reaction
    (d) all of the above

22. Ammonia (NH_3) and oxygen (O_2) can be made to react to   ( )
form only nitrogen and water. The number of moles of oxygen
consumed for each mole of nitrogen formed is:
    (a) 1.5   (b) 0.67   (c) 3.0   (d) 2.0

23. Which equation most completely represents the following   ( )
reaction? On being heated, gaseous ammonia (NH_3) decomposes to
form gaseous nitrogen (N_2) and hydrogen (H_2).
    (a) $NH_3 (g) \rightarrow N_2 (g) + H_2 (g)$
    (b) $3 H_2 (g) + N_2 (g) \rightarrow 2 NH_3 (g)$
    (c) $2 NH_3 (g) \rightarrow 6 H(g) + 2 N(g)$
    (d) $2 NH_3 (g) \rightarrow 3 H_2 (g) + N_2 (g)$

24. When balanced, the following equation for the combustion of ( )
octane has which set of coefficients?

$$\underline{\quad} C_8 H_{18} (l) + \underline{\quad} O_2 (g) \rightarrow \underline{\quad} CO_2 (g) + \underline{\quad} H_2 O(g)$$

(a) 1,25,8,18 (b) 1,25/2,16,18 (c) 1,25,8,9 (d) 2,25,16,18

25. If 4.5 moles of NO_2 are allowed to react with 3.0 moles of ( )
$H_2 O$ according to the equation:

$$3 NO_2 (g) + H_2 O(l) \rightarrow 2 HNO_3 (l) + NO(g)$$

the theoretical yield of HNO_3, in moles, would be:
(a) 4.5   (b) 3/2 × 4.5   (c) 2.0   (d) 2/3 × 4.5

26. Baking soda ($NaHCO_3$) and hydrochloric acid (HCl) react ( )
according to the equation: $HCO_3^-(aq) + H^+(aq) \rightarrow H_2 O(l) + CO_2 (g)$.
At least how many moles of $NaHCO_3$ are required for the formation
of 2.5 moles of CO_2?
(a) 1.0   (b) 2.5   (c) 5.0   (d) some other number

27. How many grams of baking soda (FW = 84.0) would be ( )
consumed in forming 10.0 grams of carbon dioxide (MW = 44)
according to the equation given in (26)?
(a) $\dfrac{10.0}{44}$ (b) $\dfrac{44 \times 84.0}{10.0}$ (c) $\dfrac{10.0 \times 84.0}{44}$ (d) $\dfrac{10.0 \times 61.0}{44}$

28. When balanced using the smallest whole numbers possible, ( )
the coefficient for Cl^- is:

$$\underline{\quad} Cl_2 (aq) + \underline{\quad} OH^-(aq) \rightarrow \underline{\quad} Cl^-(aq) + \underline{\quad} ClO_3^-(aq) + \underline{\quad} H_2 O(l)$$

(a) 1   (b) 3   (c) 4   (d) 5

29. When a certain compound of nitrogen and hydrogen reacts ( )
with oxygen, nitric oxide (NO) and water are formed. Under reaction
conditions, all four substances are gases and are found to react in a
volume ratio of 4:5:4:6, respectively. What is the simplest formula
for the nitrogen-hydrogen compound?
(a) NH_2 (b) $N_2 H_4$ (c) HN_3 (d) NH_3

30. In carrying out the reaction, ( )

$$2 \, NaHCO_3 \, (s) \rightarrow Na_2 CO_3 \, (s) + H_2 O(g) + CO_2 \, (g)$$

a student obtains an 80% yield. How many moles of $NaHCO_3$ must she have started with if her yield was 1.6 moles of $Na_2 CO_3$?
      (a) 4.0   (b) 3.2   (c) 2.6   (d) 2.0

31. When an excess of one reactant is used: ( )
    (a) more product may form than with no excess
    (b) reaction may proceed at a faster rate
    (c) less limiting reactant may remain unconsumed
    (d) all of the above

SELF-TEST ANSWERS

  1. F   (Atom for atom, Sc is heavier than Cl; actual % Cl = 70.)
  2. F   (The formula denotes only that there are two H atoms and two O atoms in a molecule of hydrogen peroxide.)
  3. T
  4. T
  5. T   (These answers are based on material of Chapters 1, 13, and 14 as well as on actual laboratory experience.)
  6. T
  7. F
  8. c   (Simply add the atomic weights for all atoms: see Chapter 2.)
  9. b
10. a
11. c   (Note that combining weights and weight percentages provide the same information.)
12. d   (Any one or all of these might work.)
13. a   (NH_2 would not give the correct number of atoms in a molecule of hydrazine, as indicated by the molecular formula $N_2 H_4$.)
14. c
15. c   (All of these formulas correspond to a MW of 90; but only (c) has the atom ratio given also by the simplest formula.)
16. a   (The largest fraction of a mole, 1.0/16.)
17. c
18. b   (Don't forget to convert to grams!)
19. a   (Mass per molecule = 9×10^5 g/6×10^{23} molecules.)
20. a
21. c   (Note that the information in (a) and (b) can be gotten from this choice, once the equation is balanced.)

22. a (You need to know that nitrogen is N_2 and to balance the equation.)

23. d

24. d

25. d $\Big($ Note that NO_2 is the limiting reactant;

$$4.5 \text{ moles } NO_2 \times \frac{2 \text{ moles } HNO_3}{3 \text{ moles } NO_2}.\Big)$$

26. b (Each mole of HCO_3^- reacting requires the use of one mole of $NaHCO_3$.)

27. c

28. d (Difficult? A systematic approach to balancing such a redox equation is given in Chapter 20. The respective coefficients are: 3, 6, 5, 1, 3.)

29. d (Recall Avogadro's hypothesis — Chapter 2: Volume ratios are equivalent to mole ratios for gases reacting under a given set of conditions. The equation:

$$4 \, NH_3 \, (g) + 5 \, O_2 \, (g) \rightarrow 4 \, NO(g) + 6 \, H_2 O(g).)$$

30. a (For 80% yield, every 2 moles $NaHCO_3$ gives 0.8 mole $Na_2 CO_3$.)

31. d (Choices (a) and (c) are discussed in Chapter 13; (b), in Chapter 14.)

SELECTED READINGS

Problem solving requires practice. For a good selection of exercises use any one of the problem or programmed manuals listed in the Preface to this guide. (Also, work as many of the problems in the text as you can.)

Copley, G. N., Linear Algebra of Chemical Formulas and Equations, *Chemistry* (October 1968), pp. 22–27.
  A general method for balancing equations by the use of determinants. Recommended only for the mathematically inclined, this is one of a series of articles.
Kieffer, W. F., *The Mole Concept in Chemistry*, New York, Van Nostrand, 1973.
  A few exercises, most of them worked out, are provided in this general discussion of stoichiometry. Will be useful later in the course too.
Strong, L. E., Balancing Chemical Equations, *Chemistry* (January 1974), pp. 13–15.
  An excellent discussion of what textbook authors and teachers usually leave to trial and error learning and "intuition."

4

Thermochemistry

QUESTIONS TO GUIDE YOUR STUDY

1. How many chemical reactions going on in the world around us can you think of in which the energy change plays an important role? (What "drives" an automobile engine; a mountain climber?)

2. What kinds of energy may be transferred during chemical reactions?

3. What is the source of the energy involved in a reaction? What happens to it? Can this energy be rationalized in terms of what happens to the atoms?

4. Is the energy transferred during a reaction quantitatively related to the reacting masses? Can you cite a simple example?

5. How is information concerning energy transfer expressed within the chemical equation? Does your interpretation of the energy transfer depend on how you write the equation for a reaction?

6. Does the energy transfer depend on the conditions under which a reaction is carried out? If so, how?

7. How do you measure the energy transfer for any given reaction? Can it be calculated for a reaction, without ever actually carrying out that particular reaction?

8. What do the thermochemical principles allow you to say about practical problems, such as the relative merits of two different fuels?

9. What are some of the fundamental problems in supplying the energy needed by modern society today and tomorrow? Are there known limitations or perhaps untapped possibilities the chemist can describe?

10. What is thermal pollution? What can be done to minimize it?

11.

12.

YOU WILL NEED TO KNOW

Concepts

1. A general notion as to the meaning of energy and the means by which it may be transferred from one object to another — see the Summary, as well as the Readings

Math

1. How to work problems in stoichiometry, and therefore how to write simple balanced equations — Chapter 3

CHAPTER SUMMARY

In every chemical reaction we observe that the mass of the products is equal to the mass of the reactants. We account for mass conservation in writing equations that are "balanced." We likewise have a bookkeeping system for energy since it too can always be accounted for. In a "thermo-chemical" equation we note how much energy is involved and show whether it is entering or leaving the system (the reaction mixture). The stoichiometry is as before — the equation is taken to represent amounts of substances in units of mole, with the accompanying energy change having the numerical value written into the equation or alongside it.

The usual way in which we measure the energies associated with chemical reactions is to measure a property of the surroundings, such as temperature, that shows how much energy has left the reaction mixture and entered the surroundings, or left the surroundings and entered the system. In all reactions, the energy is thought of as either being stored in the atoms and molecules of the system (the energy of the system increases, the sign of the energy change is positive) or being taken out of storage and passed on to the surroundings (where the energy is stored, again, in the constituent particles). What happens during a chemical reaction if the system is insulated from surrounding matter? The energy of reaction is involved in a change from one kind of storage to another. (Example: If hydrogen and oxygen explosively react to form water inside a sealed and insulated "bomb," the energy that would otherwise have been given off to the surroundings ends up being stored in the products, raising their temperature and increasing the vigor of their molecular motions.)

The transfer of energy of most concern to the chemist is that called heat. Since most reactions occur open to the atmosphere, as in test tubes, we are interested in this thermal energy transferred at constant pressure. This is called the enthalpy change, ΔH. (For most purposes, even if the pressure does change during a reaction, the heat flow is equal to ΔH.)

How do we associate energy with a molecule? In what ways can energy be stored; whence does it come? By the very fact that molecules are constantly in motion, they possess energy; the more energy, the more vigorous the motion. (What kind of evidence is there for molecular motion?) More about kinetic energy in the next chapter. And several kinds of motion may be possible for a molecule, atom, or ion: vibrating, rotating, or just moving headlong. And there's the stored "chemical energy" associated with the bonds between atoms: a result of mutual attractions, and motions, between electrons and nuclei. Differences from one molecule to the next kind of molecule in the types and numbers of these energy storing mechanisms give rise to differences in the energy transfers observed. For example, a lot more energy is normally stored in a chemical bond between the atoms in a molecule than in the motions of the molecule itself. Or again, much more energy is involved in nuclear changes than in the ordinary bond-making and bond-breaking of chemical reactions.

The fact that we can keep track of energy transfers, that energy always seems to be conserved in chemical reactions, means that we acquire the power of prediction. We can often predict the energy effect associated with a given reaction, even without first carrying out the reaction. And this is really the chief concern of this chapter — that you be able to calculate the amount of heat produced by or required for any given reaction.

But perhaps there is a more important concern for the material discussed in this chapter: that of objectively looking at our energy resources on this planet, their uses and abuses, the laws and limitations energy changes seem to follow. Just one such observation, with vast implications extending beyond chemistry, is that, as far as we can tell, heat can never be completely changed into work; there is always some unused and unusable heat left over. (In Chapter 12, we will see a rationale for the "inevitable" waste of energy in terms of the behavior of particles.) In essence, the "first law" of thermodynamics says that energy is not really consumed at all; it can only be changed in form; for example, from chemical energy into heat. The "second law" might be taken to mean that energy cannot be completely recycled; more and more, energy becomes unavailable as wasteful heat in the surroundings. How, then, do we make the best of what we've got?

BASIC SKILLS

1. Use a thermochemical equation to relate heat flow in a reaction to amounts of products or reactants.

This skill is illustrated in Example 4.1 and in the following example.

— —

For the reaction $CH_4(g) + 2O_2(g) \rightarrow CO_2(g) + 2H_2O(l)$, $\Delta H = -213$ kcal. How many grams of CH_4 must be burned to evolve one kcal of heat?_____

Here, as always in working problems dealing with balanced equations, we follow the conversion factor approach. The thermochemical equation gives us the conversion factor we need, that between grams of methane and kcal of heat evolved.

$$1 \text{ mole } CH_4 = 16.0 \text{ g } CH_4 \stackrel{\frown}{=} -213 \text{ kcal}$$

To evolve 1.00 kcal of heat, we need:

$$-1.00 \text{ kcal} \times \frac{16.0 \text{ g } CH_4}{-213 \text{ kcal}} = 0.0751 \text{ g } CH_4$$

— —

2. Calculate ΔH for a reaction from heats of formation of compounds (Table 4.1).

This skill is illustrated in Examples 4.2 and 4.3a. Note that in this type of calculation, the elements in their stable state at 25°C and 1 atm are taken as the zero of enthalpy and hence are assigned a "heat of formation" of zero.

Many of the problems at the end of Chapter 4 require the use of a table of heats of formation. Problem 4.5 is entirely analogous to Example 4.2. In Problem 4.17b, this skill must be combined with (1) above to obtain ΔH per gram of acetylene; you must also, of course, write a balanced equation for the reaction.

3. Calculate ΔH for a reaction from heats of bond formation (Table 4.2).

A typical example of this type of calculation is shown in Example 4.3b. Note that here, in contrast to skill (2) above, isolated atoms such as Cl and F are taken as the zero of enthalpy. Hence, when you work with a table of heats of bond formation, bonded species such as Cl_2 and F_2 must be included in calculating ΔH.

Problems 4.6 and 4.18 give you a chance to practice this skill. To work certain parts of these problems, you need to know how the atoms in a molecule are arranged. In 4.6c, for example, you must realize that in NH_3 and NH_2Cl, three atoms are bonded directly to the central nitrogen atom.

4. Use calorimetric data to obtain heats of reaction.

When 1.00 gram of NaOH is dissolved in 50.0 grams of water in a coffee-cup calorimeter, the temperature rises from 20.0°C to 25.0°C. What is Q for the solution process? (Take the specific heats of water and NaOH to be 1.00 cal/g°C and 0.48 cal/g°C respectively; neglect any absorption of heat by the calorimeter.) _____

We find the heat absorbed by the water:

$$50.0 \text{ g} \times (25.0°C - 20.0°C) \times 1.00 \frac{\text{cal}}{\text{g}°C} = 250 \text{ cal}$$

and by the NaOH:

$$1.00 \text{ g} \times (25.0°C - 20.0°C) \times 0.48 \frac{\text{cal}}{\text{g}°C} = 2.4 \text{ cal}$$

Adding, we find that a total of 252 cal of heat is absorbed by the contents of the calorimeter. An equal amount of heat must be given off in the solution process, i.e., $Q_{soln} = -252$ cal.

Example 4.4 is analogous to the one just worked, except that an additional step is required, the calculation of the *molar* heat of solution of NaOH. A slightly different type of "calorimeter problem" is illustrated in Example 4.5. Here, we are dealing with a bomb calorimeter, which has an appreciable heat capacity that cannot be neglected in calculating Q. (Strictly speaking, in Example 4.5 we should make a correction for the heat absorbed by the reactants, as we did in Example 4.4. However, you can readily show that such a correction would be negligible here since the amount of N_2H_4 is so small compared to the amount of water.)

Problems 4.7 and 4.19 apply the principles discussed above to data obtained from a coffee-cup calorimeter of negligible heat capacity. Problems 4.9 and 4.21 are analogous, except that a bomb calorimeter is used. Problems 4.8 and 4.20 illustrate two different experimental methods of obtaining the heat capacity of a calorimeter.

*5. Use Hess' Law to calculate ΔH for a reaction, given ΔH for other, appropriate reactions.

The general principle here is that if two or more chemical equations can be combined algebraically to give a single, overall equation, ΔH for that equation is the sum of the ΔH's for the individual equations. Symbolically, if:

$$
\begin{aligned}
&(1) \quad A + \;\; B \to C \; ; \Delta H_1 \\
&(2) \quad \underline{C + \;\; B \to D} \; ; \Delta H_2 \\
&(3) \quad A + 2B \to D \; ; \Delta H_3
\end{aligned}
$$

Then Hess' Law requires that $\Delta H_3 = \Delta H_1 + \Delta H_2$. Hence, if ΔH_1 and ΔH_2 are known, ΔH_3 can be obtained. Conversely, if ΔH_1 and ΔH_3 are given, ΔH_2 can be calculated.

This principle is applied on p. 68 of the text to obtain the heat of formation of $SnCl_4$. Many of the heats of formation listed in Table 4.1 were obtained by algebraic manipulations of this sort. Problems 4.4 and 4.16 can be solved in this way if desired; in practice, it is probably simpler to work with heats of formation directly, particularly in 4.16.

6. Use Equation 4.28 to obtain maximum efficiencies of heat engines.

— —

What is the maximum efficiency of an engine which absorbs heat at 600°C and rejects it at 400°C? _____

$$
\text{Converting to } {}^\circ K: T_a = 600^\circ + 273^\circ = 873^\circ K
$$
$$
T_r = 400^\circ + 273^\circ = 673^\circ K
$$

$$
\text{Maximum efficiency} = \frac{873 - 673}{873} = \frac{200}{873} = 0.229 = 22.9\%
$$

In other words, an engine operating between these two temperatures would convert less than 23% of the heat supplied to it into useful work; more than 77% would be wasted, dissipated to the surroundings.

— —

Problems 4.11 and 4.23 show two practical applications of this relationship. Note the advantage of using superheated steam as compared to steam at 100°C (Problem 4.11) and the very low efficiency of a "solar engine" (Problem 4.23).

Problems

1. For the reaction C_3H_8 (g) + $5O_2$ (g) → $3CO_2$ (g) + $4H_2O(l)$, $\Delta H = -531$ kcal. Calculate ΔH for:

    a) the combustion of one gram of C_3H_8.
    b) the formation of two moles of CO_2.
    c) the formation of 16.0 g of H_2O.

2. Using the data in Table 4.1 of the text, calculate ΔH for

    a) CH_4 (g) + $2Cl_2$ (g) → CCl_4 (g) + $2H_2$ (g); $\Delta Hvap$ CCl_4 = +7.0 kcal
    b) $CH_3OH(l)$ + O_2 (g) → CO(g) + $2H_2O(g)$

3. Using Table 4.2 of the text, calculate ΔH for Problem 2a above. Note that in CH_4 and CCl_4, four atoms are bonded to a central carbon atom.

4. When 1.50 g of a certain salt is added to 40.0 g of water, the temperature drops from 20.0°C to 16.2°C. Taking the specific heats of water and the salt to be 1.00 cal/g°C and 0.20 cal/g°C respectively, calculate Q for the solution process (neglect any heat flow to the calorimeter).

*5. Given $H_2O_2(l)$ → $H_2O(l)$ + $\frac{1}{2}O_2(g)$; $\Delta H = -23.5$ kcal

    $H_2(g)$ + $\frac{1}{2}O_2(g)$ → $H_2O(l)$; $\Delta H = -68.3$ kcal

    Calculate ΔH for: $H_2O_2(l)$ + $H_2(g)$ → $2H_2O(l)$
    and: $H_2O_2(l)$ → $H_2(g)$ + $O_2(g)$

*6. Calculate the efficiency of a heat engine which absorbs heat at 80°C and rejects it at 0°C. If 50 kcal of heat is absorbed, how much useful work is done?

- - - - - - - - - - - -

7. Ethylene, C_2H_4 (g), burns in air to form CO_2 (g) and $H_2O(l)$.

    a) Write a balanced equation for this reaction.
    b) Using Table 4.1, obtain ΔH for the combustion of one mole of C_2H_4.
    c) Calculate ΔH for the combustion of one gram of C_2H_4.
    d) How many grams of C_2H_4 must be burned to give off 100 kcal of heat?

8. a) Using Table 4.1, calculate ΔH for the reaction

$$CH_4 (g) + 3Cl_2 (g) \rightarrow CHCl_3 (l) + 3HCl(g)$$

b) Using Table 4.2, calculate ΔH for the reaction

$$CH_4 (g) + 3Cl_2 (g) \rightarrow CHCl_3 (g) + 3HCl(g)$$

In both CH_4 and $CHCl_3$, carbon is bonded to four other atoms.

c) Explain why the answers in (a) and (b) differ.

9. A sample of H_2 (g) weighing 1.60 g is burned in a bomb calorimeter containing 2.10 kg of water. The temperature rises from 22.0°C to 46.3°C. Assuming that all of the heat is absorbed by the water (specific heat = 1.00 cal/g°C) and the bomb (heat capacity = 108 cal/°C), calculate

a) Q for the combustion process occurring in the bomb.
b) Q for the combustion of one mole of H_2.

10. For the reaction: $C_2 H_6 (g) + \frac{7}{2} O_2 (g) \rightarrow 2CO_2 (g) + 3H_2 O(l)$, it is found that the combustion of one gram of $C_2 H_6$ gives off 12.4 kcal of heat.

a) Calculate ΔH for the combustion of one mole of $C_2 H_6$.
b) Taking the heats of formation of CO_2 (g) and $H_2 O(l)$ to be –94.1 kcal/mole and –68.3 kcal/mole respectively, use Equation 4.16 and your answer to (a) to obtain $\Delta H_f C_2 H_6$:

– – – – – – – – – – – –

11. The principal component of natural gas is methane, CH_4. When methane burns in a Bunsen burner, it combines with the oxygen of the air to give carbon dioxide and water vapor. Suppose that the burner is adjusted so that there is a 5-fold excess of air; i.e., the amount of air admitted is five times that required for the combustion of the methane. Calculate the maximum temperature, in °C, that can be reached with the burner under these conditions. In your calculations, make the following assumptions: (a) the air and methane enter the burner at 20°C; (b) air consists of N_2 and O_2 in a 4:1 mole ratio; (c) the heat evolved in the combustion of the

methane is absorbed in raising the temperature of the "products," which include CO_2, $H_2O(g)$, and the N_2 and O_2 from the excess air used; (d) the specific heats of the products are:

CO_2 (g), 0.202 cal/g°C H_2O(g), 0.446 cal/g°C

O_2 (g), 0.209 cal/g°C N_2 (g), 0.249 cal/g°C

SELF-TEST

True or False

1. In carrying out a reaction in a test tube, a student observes   ( )
that the test tube becomes cold. He should call the reaction exothermic.

2. Given the thermochemical equation: ( )

$$UF_6(l) \rightarrow UF_6(g), \Delta H = +7.2 \text{ kcal},$$

one can be sure that at least 7.2 kcal of heat must be produced or
evolved when one mole of liquid UF_6 is evaporated.

3. Another way of writing the thermochemical equation of   ( )
question (2) would be: $UF_6(l) + 7.2 \text{ kcal} \rightarrow UF_6(g)$.

4. You would expect that the heat produced in the following   ( )
reactions would have one and the same numerical value:

$$H_2(g) + \tfrac{1}{2} O_2(g) \rightarrow H_2O(l); \quad 2 H_2(g) + O_2(g) \rightarrow 2 H_2O(l)$$

5. The difference in enthalpy between one mole of Cl_2 and two   ( )
moles of atomic chlorine, Cl, both at one atmosphere pressure and
25°C is equal in magnitude to the heat of bond formation for one
mole of Cl_2.

6. Considering the example reaction: ( )

$$C(s) + O_2(g) \rightarrow CO_2(g),$$

one can always say that the heat of combustion for any substance is
the same thing as the so-called heat of formation of that substance.

7. One would expect that ΔH for the following reaction would   ( )
be less negative than the molar heat of formation of liquid water:

$$2 H(g) + O(g) \rightarrow H_2O(l)$$

8. If a system is carried through a series of steps, each of which involves an energy transfer into or out of the system, but finally ends up being identical in every way to the initial system, the total thermal energy transfer must be zero.   ( )

9. Reactions which tend to occur of their own accord, proceeding "naturally" or spontaneously, usually involve a decrease in enthalpy.   ( )

Multiple Choice

10. The largest amount of energy is stored in which one of the following systems?   ( )
    (a) 1.0 mole $H_2O(l)$ at 100°C
    (b) 1.0 mole $H_2O(l)$ at 0°C
    (c) 1.0 mole $H_2O(g)$ at 100°C and 1.0 atmosphere
    (d) 1.0 mole $H_2O(g)$ at 100°C and 1.0 atmosphere and 1.0 mole $H_2O(l)$ at 100°C have the same amount of stored energy

11. The regulation of body temperature by perspiration can be accounted for, at least in part, by the fact that   ( )
    (a) liquid water has a negative heat of formation
    (b) liquid water is a good insulator
    (c) the phase change, liquid-to-gas, is endothermic
    (d) water absorbs and releases less heat per unit temperature change than most other substances

12. The molar heat of combustion of methane, CH_4, is reported as −213 kcal. The corresponding chemical equation is: ( )
    (a) $C(g) + 4 H(g) \rightarrow CH_4 (g)$ $\Delta H = -213$ kcal
    (b) $C(s) + 2 H_2 (g) \rightarrow CH_4 (g)$ $\Delta H = -213$ kcal
    (c) $CH_4 (g) + \frac{3}{2} O_2 (g) \rightarrow CO(g) + 2 H_2 O(l)$ $\Delta H = -213$ kcal
    (d) $CH_4 (g) + 2 O_2 (g) \rightarrow CO_2 (g) + 2 H_2 O(l)$ $\Delta H = -213$ kcal

13. One can often confidently calculate the enthalpy change for a reaction before carrying out the reaction because   ( )
    (a) all heats of reaction have already been measured and tabulated
    (b) heats of formation for all known compounds have been measured
    (c) the given reaction, and its enthalpy change, may be related algebraically to other reactions already carried out
    (d) since enthalpy change depends on amount of substance reacting, one can always adjust the amount of materials so as to observe an enthalpy change of any value

14. For the reaction: $2 HCl(g) \rightarrow H_2(g) + Cl_2(g)$, $\Delta H = +44.2$ ( )
kcal. This means that:

 (a) if the given reaction is to be carried out at constant
  temperature, then the reaction mixture must be heated
 (b) the chemical bonds in the products are weaker than those
  in the reactants
 (c) $HCl(g)$ has a negative heat of formation
 (d) all of the above

15. When water evaporates at constant pressure, the *sign* of the ( )
heat effect

 (a) is negative
 (b) is positive
 (c) depends on the temperature
 (d) depends on the container volume

16. Given $\Delta H = -143.8$ kcal for the reaction: $Mg(s) + \frac{1}{2} O_2(g) \rightarrow MgO(s)$,
what would you expect to happen if the reaction were allowed to pro-
ceed at constant pressure in such a way that no heat transfer could take
place between the reaction mixture and the surroundings?

 (a) no reaction could occur
 (b) the temperature of the reaction mixture would increase
 (c) the temperature of the reaction mixture would decrease
 (d) insufficient information is given

17. When 1.00 gram of ammonia, NH_3 (FW = 17.0), is produced ( )
from N_2 and H_2 at a constant temperature (25°C) and pressure (1
atm), 648 calories are produced. The molar heat of formation of
ammonia, in kilocalories, is

 (a) −0.648 (17.0)   (b) 17.0/648
 (c) −0.648/17.0   (d) +0.648 (17.0)

18. Which one of the following reactions would you expect to be ( )
the source of the largest amount of heat?

 (a) $CH_4(l) + 2 O_2(g) \rightarrow CO_2(g) + 2 H_2O(g)$
 (b) $CH_4(g) + 2 O_2(g) \rightarrow CO_2(g) + 2 H_2O(g)$
 (c) $CH_4(g) + 2 O_2(g) \rightarrow CO_2(g) + 2 H_2O(l)$
 (d) $CH_4(g) + \frac{3}{2} O_2(g) \rightarrow CO(g) + 2 H_2O(l)$

19. In what order would you arrange the following reactions so ( )
that the magnitude of the enthalpy change increased in that order?

   A. $^{238}U + n \rightarrow {}^{239}U$
   B. $H_2O(l) \rightarrow H_2O(s)$
   C. $C(coal) + O_2(g) \rightarrow CO_2(g)$

 (a) ABC   (b) BCA   (c) CAB   (d) ACB

20. Given the heats of combustion for diamond (−94.50 kcal) ( )
and graphite (−94.05 kcal), both composed of pure carbon, would
you expect the formation of diamond from graphite to be
endothermic or exothermic?
     (a) endothermic   (b) exothermic
     (c) depends on the temperature   (d) insufficient information

21. Hydrogen, H_2, has been proposed as a fuel of the future. ( )
Knowing that very little hydrogen exists in this elemental state, you
would expect that
     (a) hydrogen could play no role in the transfer of energy
         from one system to another
     (b) elemental hydrogen might well supplant all other fuels
     (c) energy received from another source might be stored in
         H_2 and later released during the combustion of the
         hydrogen
     (d) all of the above

22. In 1970, we obtained about 23% of our energy from coal, ( )
about 75% from oil and natural gas, about 2% from water power, and
about 0.2% from nuclear processes. Which one of these percentages
seems most likely to decrease by 2000 AD?
     (a) coal   (b) oil and natural gas
     (c) water power   (d) nuclear processes

23. The greatest potential for meeting our energy needs in the ( )
long run is offered by
     (a) coal   (b) petroleum
     (c) fission   (d) fusion on the earth or sun

24. In order to determine the heat evolved when a sample is ( )
burned in a bomb calorimeter, you must know:
     (a) the mass and specific heat of the water in the calorimeter
     (b) the temperature change
     (c) the heat capacity of the bomb
     (d) all the above

25. Given that ΔH = +44.2 kcal for the reaction ( )
2 HCl(g) → H_2 (g) + Cl_2 (g), the molar heat of formation of the H–Cl
bond
     (a) is −44.2 kcal
     (b) is −22.1 kcal
     (c) is +22.1 kcal
     (d) cannot be determined without additional information

26. The molar heats of formation of the bonds in the halogen ( )
molecules F_2, Cl_2, Br_2, and I_2 are –37 kcal, –58 kcal, –46 kcal, and
–36 kcal respectively. Of these, the strongest bond is found in
     (a) F_2 (b) Cl_2 (c) Br_2 (d) I_2

SELF-TEST ANSWERS

1. **F** (To restore the temperature to its initial value, heat must eventually be added to the test tube.)

2. **F**

3. **T**

4. **F** (The first heat effect would be that for forming one mole of water; the second, twice as much heat for twice the amount of water. This is how the equation stoichiometry is interpreted.)

5. **T**

6. **F** (Solid carbon is undergoing combustion; gaseous carbon dioxide is being formed from the elements.)

7. **F** (The given reaction can be considered as the sum of the two reactions: $H_2(g) + \frac{1}{2} O_2(g) \rightarrow H_2O(l)$; $2 H(g) + O(g) \rightarrow H_2(g) + \frac{1}{2} O_2(g)$. The second step releases additional energy, making the overall ΔH more negative.)

8. **F** (Q could be practically anything, provided other kinds of energy transfer were involved.)

9. **T** (You probably expect this from common experience: heat is generally produced by reactions. For a closer look at spontaneous reactions, see Chapter 12.)

10. **c** (ΔH is positive for $H_2O(l) \rightarrow H_2O(g)$; additional energy is stored in the gas. So, steam causes the more severe burns.)

11. **c** (Evaporation requires heat; heat is removed from the hot body.)

12. **d** (Note that CO_2 is generally formed, and not CO, in the reaction chemists call combustion — unless otherwise noted.)

13. **c**

14. **d**

15. **b** (Choice c, for example, is ruled out since ΔH doesn't change much with temperature.)

16. **b**

17. **a** (Check your units.)

18. **c** (The energy released increases from a to b to c; as well as from d to c, since burning CO to form CO_2 would release the larger amount of energy associated with c. Do you also see the rationale for always specifying the states?)

19. **b**

20. a   (ΔH = +0.45 kcal/mole. This is small compared to most heats of reaction for chemical changes. So why were diamonds only recently synthesized?)
21. c   (Readings.)
22. b   (Readings.)
23. d
24. d
25. d   (To find: ΔH for the reaction $H(g) + Cl(g) \rightarrow HCl(g)$.)
26. b

SELECTED READINGS

The Biosphere, San Francisco, W. H. Freeman, 1970 (the September 1970 issue of *Scientific American*).
  See particularly the chapters on the energy cycles of the earth and on human energy production.
Daniels, F., *Direct Use of the Sun's Energy,* New Haven, Yale, 1964.
  A pioneering and comprehensive work on solar energy; illustrated, almost "how to do it."
Faraday, M., *The Chemical History of a Candle,* New York, Viking Press, 1960.
  A delightful exploration of many aspects of combustion; dates back to 1860.
Holdren, J. and P. Herrera, *Energy,* New York, Sierra Club, 1971.
  A readable, apparently unbiased account of present and likely future sources, uses and abuses of energy. Includes "case studies" of citizen action groups.
Hydrogen: Likely Fuel of the Future, *C & EN* (June 26, 1972), pp. 14-17.
  The first in a three-part series describing "a new energy system" employing hydrogen as a carrier (not a primary source) of energy; sources of hydrogen; implications of large scale use.
Mahan, B. H., *Elementary Chemical Thermodynamics,* New York, W. A. Benjamin, 1963.
  A good introduction to classical thermodynamics, going somewhat farther and deeper into the subject than this text. (Also see the paperback below and other references listed for Chapter 12.)
Pimentel, G. C. and R. D. Spratley, *Understanding Chemical Thermodynamics,* San Francisco, Holden-Day, 1969.
  With humor and lots of commonsense illustrations and applications, the authors introduce you to most of the uses and meanings you are likely to want, in this course and elsewhere.
Sandfort, J. F., *Heat Engines: Thermodynamics in Theory and Practice,* Garden City, New York, Doubleday-Anchor, 1962.
  A very easily read introduction to engineering thermodynamics and its history.
Strong, L. E. and W. J. Stratton, *Chemical Energy,* New York, Reinhold, 1965.
  A very carefully developed discussion — though not easy reading — on the same level as the text, with an emphasis on interpreting thermodynamics in terms of molecular properties (structure and bonding; disorder).

For further practice in the calculations of thermochemistry, see Barrow, and the other problem manuals listed in the preface.

5

The Physical Behavior of Gases

QUESTIONS TO GUIDE YOUR STUDY

1. What materials can you think of that usually exist as gases? What properties do they share?

2. What conditions favor the existence of a substance as a gas rather than as a liquid or a solid?

3. How do you measure properties of gases such as temperature and pressure? How would you weigh a sample of gas?

4. Is there a simple relationship among the properties of a gas that generally holds for all gases? Can you rationalize such a relationship in terms of atomic-molecular theory? (For example: how does the behavior of molecules explain the relation between temperature and pressure for the air inside a tire?)

5. How do mixtures of gases behave? How is their behavior related to that of a pure gaseous substance?

6. Can the quantities of gases participating in a chemical reaction be simply expressed in terms of weights; moles; volumes?

7. How does the volume of gaseous reactant or product depend on reaction conditions?

8. What are some of the practical applications (as well as support for other chemical principles) of our knowledge of gaseous behavior?

9. What experimental support is there for our ideas about the nature of the molecules in a gas?

10. How do you account for the differences in properties among gases and between gases and liquids and solids?

11.

12.

YOU WILL NEED TO KNOW

Concepts

1. Meaning of molecular weight and how to calculate it from a formula — Chapter 2

Math

1. How to solve first and second order equations for any one variable

Examples: Rewrite the equation $PV = gRT/M$ in the form $M = \ldots$

Solve the equation $\frac{1}{3} Mu^2 = RT$ for u.

2. How to find the square root of a number (most simply done by using a slide rule or a table of logarithms) — see readings listed in the Preface.

CHAPTER SUMMARY

The physical behavior of gases is described concisely by the Ideal Gas Law, $PV = nRT$, which forms the central theme of this chapter. This equation tells us how the four experimental variables, pressure, volume, number of moles, and absolute temperature, are related to one another. We see, for example, that for a given sample of gas at a fixed temperature (i.e., n and T constant): $PV = \text{constant}$, or $P = \text{constant}/V$. That is, the ideal gas law embodies Boyle's Law.

Again, we are reminded that for a given sample of gas at a fixed pressure (i.e., P and n constant): $V = \text{constant} \times T$ (Law of Charles and Gay-Lussac). In still another case, we see that when temperature and pressure are held constant: $V = \text{constant} \times n$, which implies Avogadro's Law (equal volumes of all gases at the same temperature and pressure contain equal numbers of molecules).

Frequently, we use the Ideal Gas Law to calculate one variable given the values of the others. In order to carry out such calculations, we must know the magnitude of the gas law constant, R, in the proper units. For our purposes in this chapter, it is most convenient to express R in liter-atmosphere/mole-degree K: $R = 0.0821 \text{ lit atm/mole}°K$. From time to time, particularly in later chapters of the text, we will use R in other units: $R = 1.99 \text{ cal/mole }°K = 8.31 \times 10^7 \text{ ergs/mole }°K$. Note that R always has the units of energy/mole$°K$. (A liter atmosphere represents the work done when a piston sweeps through a volume of one liter against a constant pressure of one atmosphere.)

For certain applications of the Ideal Gas Law, it is convenient to substitute for the number of moles, n, its equivalent in grams, i.e., n = g/M, where g is the number of grams of the gas and M is its gram molecular weight. The resulting equation, $PV = gRT/M$, can be used to determine the molecular weight of a gas from measured values of P, V, g, and T. Furthermore, recognizing that the density is mass divided by volume, we can obtain a relation which tells us, among other things, that the densities of different gases at the same temperature and pressure are in the same ratio as their molecular weights.

The Ideal Gas Law can be applied to gas mixtures as well as to pure gases. We can, for example, write: $P_A V = n_A RT$, where P_A is the partial pressure and n_A the number of moles of gas A in the mixture. By combining similar expressions for each gas in the mixture, it is possible to obtain Dalton's Law. This law is particularly useful in making calculations involving "wet" gases. Such mixtures, in which water vapor is one component, are commonly formed in the laboratory when gases are collected over water.

The validity of the Ideal Gas Law is confirmed by experimental measurements which require no assumptions about the behavior of gas molecules. However, by making some rather simple assumptions concerning molecular motion, embodied in what is known as kinetic molecular theory, it is possible to derive the law from "first principles." This exercise in logic was one of the great triumphs of 19th century science; indeed, it provided the first really convincing evidence for the existence of molecules and atoms. A key postulate of kinetic theory is that, at a given temperature, molecules of all gases have the same kinetic energy of translation. Specifically, $\epsilon = mu^2/2 = \text{constant} \times T$, where m is the mass and u the average velocity of a molecule. Starting with this postulate, it is possible to derive equations for the relative rates of effusion of different molecules (Graham's Law) or the average molecular velocity of a particular gas at a given temperature. You should keep in mind that u in these equations is an *average* velocity; at any given instant, virtually all of the molecules in a gas sample are moving at velocities either greater or smaller than u. This statistical range or distribution of velocities results, as we intuitively expect, from collisions between molecules: a collision between two molecules will change the velocities of the two particles, in terms of magnitude, generally, as well as direction.

We will see in later chapters (particularly 9 and 14) how important is the existence of the distribution of energies among molecules. The fact that some molecules have considerably greater than average energy will be important in our interpretation of such seemingly diverse phenomena as the evaporation of a liquid and the way in which reaction rate changes with temperature.

The simple kinetic molecular model of gases ignores interactions between molecules and assumes their volume to be negligible in comparison to that of their container. At low pressures and high temperatures, where the molecules are far apart, these approximations are justified and we find that the Ideal Gas Law describes very well the behavior of real gases. However, at high pressures and low temperatures, intermolecular forces and molecular volumes become significant and real gases deviate considerably from ideal behavior. Indeed, if the molecules of a gas approach each other closely enough, condensation to a liquid or solid occurs, in which case the Ideal Gas Law is inapplicable.

BASIC SKILLS

You will find that the material in this chapter can best be mastered by working problems. In particular, you should be able to perform the following skills.

1. Use the Ideal Gas Law to:
   a) Determine the effect of a change in conditions (e.g., a change in T or P) upon a particular variable (e.g., V).

- -

A sample of gas has a volume of 312 ml at 273°K and 760 mm Hg. What volume will the same sample occupy at 298°K and 740 mm Hg? _____

For the gas in both states: PV = nRT. But, from the statement of the problem, the number of moles of gas, n, remains constant. The gas constant, R, must of course remain unchanged. It follows that the quantity PV/T must have the same value in the final and initial states. That is:

$$nR = \frac{P_2 V_2}{T_2} = \frac{P_1 V_1}{T_1}$$

where the subscript 1 refers to the initial state and 2 to the final state. Solving for the variable we need, V_2,

$$V_2 = \frac{P_1 V_1}{T_1}\left(\frac{T_2}{P_2}\right) = V_1 \times \frac{T_2}{T_1} \times \frac{P_1}{P_2}$$

Having derived the relationship required, all that remains is to substitute numbers. From the statement of the problem, V_1 = 312 ml, T_2 = 298°K, T_1 = 273°K, P_1 = 760 mm Hg, P_2 = 740 mm Hg.

$$V_2 = 312 \text{ ml} \times \frac{298°K}{273°K} \times \frac{760 \text{ mm Hg}}{740 \text{ mm Hg}} = 350 \text{ ml}$$

- -

This technique is further illustrated in Examples 5.1 and 5.2. Note that we can always derive the relationship required, which may involve two variables (e.g., P and V in Example 5.1, V and T in Example 5.2) or three variables (V, T, and P in the example above), from the Ideal Gas Law by applying simple algebra. In case you are tempted to try to memorize these relationships, we should point out that there are 10 of them! Notice that in order to solve problems of this type, both T_2 and T_1 must be in °K. Furthermore, P_2 and P_1 (or V_2 and V_1) must be expressed in the same units. We could, for example, express both P_2 and P_1 in atmospheres or both in mm Hg, but we could not use P_2 in atm and P_1 in mm Hg.

Problems 5.4, 5.8, 5.9, 5.23 and 5.27 are entirely analogous to the examples cited. Problem 5.7 introduces a minor complication (gauge pressure); Problem 5.26 is probably the most difficult of this type (and the most interesting!).

b) Solve for one variable (e.g., V) given the values of the other three (e.g., n, P, and T).

- -

What volume is occupied by 2.10 moles of an ideal gas at 20°C and 1.50 atm? _____
Solving the Ideal Gas Law for V:

$$V = \frac{nRT}{P}$$

Substituting n = 2.10 moles; R = 0.0821 lit atm/mole°K; T = (20 + 273)°K = 293°K; P = 1.50 atm:

$$V = \frac{(2.10 \text{ moles}) (0.0821 \text{ lit atm/mole°K}) (293°K)}{1.50 \text{ atm}} = 33.7 \text{ lit}$$

- -

Example 5.3 is analogous except that you must first convert grams of XeF_4 to moles. Example 5.4 is slightly more complex, requiring the conversion of moles to pounds in the final step. Of the problems of this type at the end of Chapter 5, the simplest is 5.28.

Note that in this application of the Ideal Gas Law, the units used for the variables must be consistent with the value chosen for R. It is probably simplest to stay with a single value of R (0.0821 lit atm/mole°K) and, if necessary, convert P to atmospheres, as in Example 5.4, or V to liters.

This skill can be combined with the principles regarding mass relations in chemical reactions (Chapter 3) to relate the number of grams or moles of one species to the volume of a gaseous reactant or product. This application is shown in Example 5.7. A somewhat simpler example is presented on the following page.

- -

In the reaction $CH_4(g) + 2O_2(g) \rightarrow CO_2(g) + 2H_2O(l)$, what volume of O_2, measured at 20°C and 1.02 atm, is required to react with 4.38 moles of CH_4? _____

Noting that the coefficients of the balanced equation give us the mole ratio in which CH_4 and O_2 react, we deduce that $2 \times 4.38 = 8.76$ moles of O_2 are required. The volume of O_2 can now be calculated exactly as in the previous example:

$$V = \frac{nRT}{P} = \frac{(8.76 \text{ moles})(0.0821 \text{ lit atm/mole}°K)(293°K)}{1.02 \text{ atm}} = 207 \text{ lit}$$

- -

Problems 5.12, 5.15, and 5.34 are of this type. Note that in 5.12 and 5.34, a correction must first be made for the partial pressure of water vapor.

c) Calculate the density of a gas at a given temperature and pressure.

Here, and for d) below, it is convenient to write the Ideal Gas Law in the form:

$$PV = gRT/M \qquad \text{(Equation 5.7)}$$

Realizing that density is the ratio of mass (g) to volume (V), we obtain the general relation:

$$\text{density} = \frac{g}{V} = \frac{MP}{RT}$$

The use of this relation is illustrated in Example 5.5 and Problem 5.10. Problem 5.29 is a little more subtle in that you first have to calculate an effective molecular weight for air.

d) Calculate the molecular weight of a gas, knowing the mass of a given volume (or the density) at a known P and T.

See Example 5.6 and Problems 5.11 and 5.30.

2. Relate the volumes of gases (measured at the same P and T) in a chemical reaction.

The basic principle here is a simple one: the coefficients of a balanced equation relate not only the numbers of moles of reactants and products, but also the volumes of gases measured at the same temperature and pressure. To illustrate this principle, consider the second example shown under Skill 1b) above. Having deduced that 207 liters of O_2 are required, we conclude that one half that volume, or 104 liters, of CO_2 would be produced, if both gases are measured at 20°C and 1.02 atm. Problems 5.13 and 5.32 can be solved in the same manner.

A note of caution: this simple relationship between reacting volumes is restricted to *gaseous* reactants or products at the *same temperature and pressure*. In the example just cited, we could not have taken the volume of liquid water to be equal to that of O_2. Neither could we have assumed a 2:1 volume ratio for O_2 and CO_2 if one of the gases were at a different temperature or pressure from the other.

3. Use Dalton's Law to obtain partial pressures of gases in mixtures.

A sample of H_2 is collected over water at a total pressure of 742 mm Hg. The partial pressure of water vapor in the gaseous mixture is 21 mm Hg. What is the partial pressure of H_2 ? _____

Applying Dalton's Law:

$$P_{total} = P_{H_2} + P_{H_2O}$$

$$P_{H_2} = P_{total} - P_{H_2O} = 742 \text{ mm Hg} - 21 \text{ mm Hg} = 721 \text{ mm Hg}$$

This simple operation is often the first step in a gas law calculation. In Example 5.9, we first obtain the partial pressure of H_2 and then use the Ideal Gas Law to obtain the number of moles of H_2. Problems 5.31 and 5.34 are solved similarly. In Problem 5.12 you calculate first the partial pressure of O_2 and then the number of moles of O_2, and finally use the balanced equation to obtain grams of $KClO_3$.

4. Use Graham's Law to relate the molecular weights of two gases to their rates of effusion.

It is found that the rate of effusion of a certain gas is 0.600 times that of O_2 under the same conditions. What is the molecular weight of the gas? _____

Applying Graham's Law:

$$\frac{r_X}{r_{O_2}} = \left(\frac{M_{O_2}}{M_X}\right)^{1/2}$$

Now, since $r_X = 0.600 \ r_{O_2}$, we see that the ratio r_X/r_{O_2} is 0.600. The molecular weight of O_2 is 32.0. Hence:

$$0.600 = \left(\frac{32.0}{M_X}\right)^{1/2}$$

Squaring both sides and solving for M_X:

$$0.360 = \frac{32.0}{M_X} \; ; \; M_X = \frac{32.0}{0.360} = 88.9$$

- -

In many problems of this type, the times of effusion rather than the rates are specified (Example 5.10; Problem 5.19). The time required for an event to take place is inversely related to the rate at which it occurs.

5. Calculate the average speed of a particular molecule at a given temperature.

This calculation is illustrated in Example 5.11 and in Problem 5.36. Note that the value of R required here is 8.31×10^7 ergs/mole°K. Since an erg has the dimensions of g cm^2/sec^2, the calculated speed will be in cm/sec.

Problems

1. a) A sample of a gas occupying a volume of 200 cm^3 at 30°C exerts a pressure of 512 mm Hg. What will be the pressure of the gas if:
    — the volume is reduced to 152 cm^3 ?
    — the temperature is increased to 60°C?
    — the volume increases to 210 cm^3 and the temperature drops to 0°C?
    — both the number of moles of gas and the volume are doubled?
    (Assume in each case that all variables not specified are held constant.)
   b) How many moles of gas are there in the sample described in (a)? If the gas is N_2, how many grams are present?
   c) What is the density of N_2 gas at 25°C and 750 mm Hg?
   d) What is the molecular weight of a gas which has a density of 1.65 g/liter at 100°C and 1.00 atm?

2. In the reaction: $4NH_3$ (g) + $7O_2$ (g) → $4NO_2$ (g) + $6H_2O$(l), 13.0 liters of NH_3 are consumed. What volume of O_2, measured at the same temperature and pressure, is required? What, if anything, can you say about the volume of NO_2 produced? the volume of H_2O?

3. A sample of oxygen saturated with water vapor at 30°C exerts a total pressure of 680 mm Hg. Using the table on p. 659 to obtain the partial pressure of water, calculate the partial pressure of O_2.

4. What is the ratio of the rate of effusion of SO_2 to that of O_2 at the same T and P?

*5. What is the average speed of an N_2 molecule at 25°C?

- - - - - - - - - - - -

6. A flask full of air at 20°C is heated at constant pressure to 100°C. What fraction of the gas molecules remains in the flask? What fraction escapes?

7. What pressure is exerted by 20.0 g of N_2 occupying a volume of 2.61 liters at 25°C?

8. The reaction $2NO(g) + O_2(g) \rightarrow 2NO_2(g)$ produces 12.0 liters of NO_2 measured at 25°C and 1.00 atm.

    a) How many moles of NO_2 are produced?
    b) How many moles of NO are required? how many grams of NO?
    c) What volume of NO at 25°C and 1.00 atm is required?
    d) What volume of NO at 0°C and 2.60 atm is required?

9. It is found that a certain gas occupying a 225 ml flask at 100°C and 750 mm Hg weighs 0.460 g.

    a) What is the molecular weight of the gas?
    b) What would be the density of this gas at 0°C and 1.00 atm?

10. A mixture of 1.50 moles of N_2 and 0.60 mole of He is confined in a 25.0 liter vessel at 50°C.

    a) What is the partial pressure of N_2? of He?
    b) What is the total pressure of the mixture?

11. It takes 12.0 seconds for a certain number of moles of O_2 to effuse through a small orifice. How long will it take for the same number of moles of N_2 to effuse under the same conditions?

- - - - - - - - - - - -

12. A 0.680 gram sample of phosphine, PH_3 (M = 34.0), occupies a volume of 589 ml at 77.0°C and 742 mm Hg. When completely decomposed to the elements at this temperature and pressure, the measured volume is 1030 ml.

    a) Calculate the total number of moles of gaseous products.
    b) Based on the information given and the fact that gaseous hydrogen is H_2, write the balanced equation for the reaction. Be sure that the equation does indeed fit the data!

SELF-TEST

True or False

1. To prepare pure nitrogen gas from a sample of air, one would   (   )
probably employ fractional distillation.

2. The molecules of a sample of any gas, under most conditions,   (   )
can be characterized as being separated by relatively large distances.

3. At a temperature of 20°C, about 3.4 grams of carbon dioxide   (   )
gas will dissolve in a liter of water under a pressure of 1.0 atmosphere. At a higher temperature, but at the same pressure, one would
expect that more carbon dioxide will dissolve.

4. The pressure exerted by one gram of O_2 in a ten liter   (   )
container at 27°C will be: 1.00(0.0821)(300)/10.0 atmospheres.

5. In a mixture of two gases, with a total pressure of 760 mm   (   )
Hg, the partial pressure of chlorine gas is 380 mm Hg. This means
that half of the molecules in the sample are chlorine.

6. The average kinetic energy of an oxygen molecule and the   (   )
average kinetic energy of a hydrogen molecule, both gases at the
same temperature, are in the same ratio as their molecular weights.

7. The average speed of a gas molecule depends only on the   (   )
absolute temperature.

8. Two separate samples of the same gaseous substance at the   (   )
same pressure would have densities in the same ratio as their absolute
temperatures.

9. The fact that a sample of gas would not have zero volume at   (   )
the absolute zero of temperature is a consequence of the fact that
absolute zero cannot be reached.

Multiple Choice

10. The chemical analysis of a mixture of gases is most likely to   (   )
involve

    (a) density measurements
    (b) mass spectrometry
    (c) boiling point measurements
    (d) fractional crystallization

11. Which one of the following substances would you expect to ( )
normally exist as a gas at room temperature?
(a) CH_4 (b) C_5H_{12} (c) C_6H_6 (d) C_3H_7OH

12. The volume of a mole of gas is ( )
(a) 22.4 liters
(b) directly proportional to pressure and absolute temperature
(c) directly proportional to pressure, inversely proportional to absolute temperature
(d) inversely proportional to pressure, directly proportional to absolute temperature

13. The inflation of an automobile tire to a pressure of "20 ( )
pounds" means:
(a) the pressure exerted by the air in the tire is 20 lb/in^2
(b) the weight of the air in the tire is 20 pounds
(c) the pressure of the air in the tire is 20 lb/in^2 higher than the air pressure outside the tire
(d) the pressure of the air inside the tire is about 5 lb/in^2

14. A certain mountain rises to 14,100 feet above sea level. The ( )
pressure at the top is about 17.7 inches (of mercury). If you blew up
a balloon at sea level, where the pressure happened to be 29.7 inches,
and carried it to the top of the mountain, by what factor would its
volume change?
(a) there would be no change   (b) $29.7 - 17.7$
(c) 29.7/17.7   (d)   17.7/29.7

15. When equal numbers of moles of two gases at the same ( )
temperature are mixed in a container, the pressure of the gaseous
mixture is
(a) given by Gay-Lussac's law of combining volumes
(b) the product of the pressures each gas would have if alone
(c) the difference of the pressures each gas would have if alone in the container
(d) the sum of the pressures each gas would have if alone

16. A gaseous substance is known which can be decomposed to ( )
give only the elements phosphorus and hydrogen. When all three
substances are gases at a convenient temperature and pressure, it is
found that four volumes of the compound give one volume of
phosphorus and six volumes of hydrogen. The simplest interpretation
is that phosphorus gas is
(a) P   (b) P_2 (c) P_3 (d) P_4

17. An equation representing a reaction which is consistent with ( )
the data of question 16 would be:
    (a) $4 PH_2 (g) \rightarrow 2 P_2 (g) + 4 H_2 (g)$
    (b) $4 PH_3 (g) \rightarrow P_4 (g) + 6 H_2 (g)$
    (c) $2 PH_3 (g) \rightarrow 2 P(g) + 3 H_2 (g)$
    (d) some other equation

18. Two flasks of equal volume are filled with different gases, A ( )
and B, at the same temperature and pressure. The weight of gas A is
0.34 gram while that of gas B is 0.48 gram. It is known that gas B is
ozone, O_3, and that gas A is one of the following. Which of the
following is most likely to be gas A?
    (a) O_2 (b) $H_2 S$ (c) SO_2 (d) cannot say

19. Which one of the following statements about the gases A and ( )
B of question 18 is false?
    (a) the numbers of molecules of A and B are equal
    (b) the average speeds of molecules A and B are the same
    (c) the masses of individual molecules of A and B compare in
        the same way as the masses of the samples
    (d) the average translational energies of molecules of A and B
        are the same

20. The fact that the ideal gas law only approximately describes ( )
the behavior of a gas can be partly explained by the idea that
    (a) R is not really a constant
    (b) gas molecules really do have zero volume
    (c) the kinetic energy of gas molecules is not really directly
        proportional to the absolute temperature
    (d) gas molecules really do interact with each other

21. Real gases behave most nearly like the ideal gas law says they ( )
do at
    (a) high temperatures, low pressures
    (b) low temperatures, high pressures
    (c) high temperatures, high pressures
    (d) low temperatures, low pressures

22. The van der Waals equation, $P = RT/(V - b) - a/V^2$, ( )
incorporates the following correction(s) to the ideal gas law in order
to account for the properties of real gases:
    (a) the possibility of chemical reaction between molecules
    (b) the finite volume of molecules
    (c) the quantum behavior of molecules
    (d) average kinetic energy is inversely proportional to
        temperature

23. To convert a sample of air into a liquid, you would probably   ( )
have to
    (a) increase the temperature and pressure of the sample
    (b) decrease the temperature and increase the pressure of the
        sample
    (c) cool it to $0°K$
    (d) the task is an impossible one

24. Your lecturer opens a bottle of hydrogen sulfide gas, H_2S,   ( )
and a bottle of gaseous diethyl ether, $C_4H_{10}O$, at the same time.
Both gases are at the same temperature and pressure. Which of the
two should you be able to smell first? (Both have characteristic
odors.)
    (a) the ether
    (b) H_2S
    (c) both at the same time
    (d) neither gas would escape into the room

25. What must be the molecular weight of a gas that effuses 1/4   ( )
as rapidly as CH_4 $(M = 16.0)$?
    (a) 4   (b) 16   (c) 64   (d) 256

26. When 6.00 liters of N_2 and 6.00 liters of H_2 are mixed, the   ( )
following reaction occurs: N_2 (g) + $3H_2$ (g) → $2NH_3$ (g). What volume
of NH_3 is produced at the same temperature and pressure at which
these volumes of reactants were measured? (Assume 100% yield.)
    (a) 2.00 liters   (b) 4.00 liters
    (c) 6.00 liters   (d) 12.0 liters

SELF-TEST ANSWERS

   1. T (The air would be liquefied and then fractionally distilled. See
        Chapter 1.)
   2. T (Large compared to molecular diameters.)
   3. F (Think of what would happen if you warmed up a bottle of
        carbonated water, or a bottle of Coke. More on solubility —
        Chap. 10.)
   4. F $(P = gRT/MV)$
   5. T
   6. F (They would be the same.)
   7. F (Also on molecular weight.)
   8. F (Inversely: density = mass/volume = PM/RT, where M = molecu-
        lar weight. So, $d(1)/d(2) = T(2)/T(1)$ for gases (1) and (2).)
   9. F (Any real gas condenses to a liquid well above absolute zero.)
 10. b

11. a   (Low molecular weight — Chap. 8.)
12. d   (The volume would be 22.4 liters only under standard conditions.)
13. c   (Now you know!)
14. c   (Larger at lower pressure.)
15. d   (Dalton's law of partial pressures.)
16. d   (See question 17, also based on Avogadro's hypothesis.)
17. b
18. b
19. b
20. d
21. a
22. b
23. b   (These are the conditions under which the real gas deviates most from the gas laws.)
24. b   (With the lower molecular weight — and given the same average translational energies — the H_2S would have the higher average speed.)
25. d
26. b

SELECTED READINGS

Conant, J. B., *Science and Common Sense*, New Haven, Yale, 1951.
> *A case history approach to science for the layman; particularly interesting and readable are the sections on Boyle's work with gases and Lavoisier's pioneering experiments on combustion.*

Hildebrand, J. H., *An Introduction to Molecular Kinetic Theory*, New York, Reinhold, 1963.
> *A more extensive discussion than the present chapter, with much more on real gases; the theory is extended to describe some of the properties of liquids and solids. The author often injects himself into the discussion.*

Neville, R. G., The Discovery of Boyle's Law, 1661-62, *J. Chem. Ed.* (July 1962), pp. 356-359.
> *Interesting mainly for the quotations from Boyle.*

Steinherz, H. A. and P. A. Redhead, Ultrahigh Vacuum, *Scientific American* (March 1962), pp. 78-90.
> *A very graphic discussion of the instruments for measuring and producing very low pressures ($< 10^{-8}$ mm Hg), as well as of its uses.*

The Electronic Structure of Atoms

QUESTIONS TO GUIDE YOUR STUDY

1. On the basis of what you have learned so far, what can be said about the properties of specific substances? (For example: at 150°C and one atm., is water a gas, liquid, or solid; is it colorless; what is its density?)

2. Are there regularities in the observed properties of the chemical elements, for example, in their molecular formulas?

3. How can the information referred to in (2) be systematized? How might it be used?

4. Are there irregularities in the properties of elements? If so, how are they accounted for?

5. To what extent can properties of substances be explained in terms of the nature of individual atoms and molecules?

6. What evidence is there for the idea that atoms are themselves composed of smaller parts? (Review Chapter 2.)

7. What experimental evidence demonstrates that there are regularities in arrangement of electrons in atoms, both within a particular kind of atom and for atoms of all the elements?

8. How do you conveniently represent the structure of an atom and the energy associated with a change in this structure? How can you describe in a meaningful way where electrons are and what they are doing?

9. Are there general, simplifying correlations between electronic structure of atoms and their properties, such as the strength of the forces between atoms in a molecule?

10. Do you suppose that modern quantum theory is the "final word"?

11.

12.

YOU WILL NEED TO KNOW

Concepts

1. General ideas of atomic and nuclear composition (kinds, numbers, and charges of nuclear particles; meaning of atomic number, mass number, isotope . . .) — Review Chapter 2

2. A general idea of what composes the electromagnetic spectrum (infrared, visible, ultraviolet, x-rays . . .) — the names of the regions and their approximate energies, wavelengths, and frequencies — See this chapter, as well as any introductory physics textbook

Math

1. No new math is required here

2. How to compute the charge of a monatomic ion, given the number of electrons and the nuclear charge (or atomic number) — Chapter 2

CHAPTER SUMMARY

The electronic structures of atoms, the major theme of this chapter, have occupied the attention of chemists and physicists for more than a century. As so often happens, theories were developed and subsequently modified to explain puzzling experimental observations. In particular, it was found that light emitted by "excited" gaseous atoms appears at certain discrete wavelengths. To explain this phenomenon, it was proposed that electrons in atoms are restricted to discrete energy levels whose separation, ΔE, is related to the wavelength, λ, of the emitted light by the Einstein equation: $\Delta E = hc/\lambda$ (for λ in Å and ΔE in ergs/atom, hc = 1.99×10^{-8}; for λ in Å and ΔE in kcal/mol, hc = 2.86×10^5). This postulate of the quantization of electronic energies is basic to modern quantum theory.

In 1913, Niels Bohr derived an equation for the allowable energy levels in a one-electron atom. This equation agreed remarkably well with the observed spectrum of hydrogen. The Bohr model predicted that the electron, in moving about the nucleus, would be restricted to discrete circular orbits of fixed radius. Unfortunately, this simple model breaks down for any species with more than one electron. Modern quantum theory tells us that exact positions of electrons cannot be specified. The best we can hope to do is to calculate the probability of finding an electron in a particular region.

Electronic energies can be calculated, at least in principle, from the Schrödinger wave equation. You were probably relieved to find that we did not attempt in this text to solve this formidable mathematical expression for allowed energies. Instead, we discussed briefly a simpler, less general expression, Equation 6.12, corresponding to the so-called "particle in a box" model. This model helps us to understand why small particles confined to very small regions of space do not obey the classical laws of motion with which we are familiar.

To completely characterize an electron in an atom, we indicate its

(1) *principal energy level,* specified by the quantum number n, which is restricted to positive, integral values (n = 1, 2, 3, . . .);

(2) *sublevel,* designated (illogically) by the letters s, p, d, f, . . . Within a principal level of quantum number n, there are n sublevels. For n = 1, we have only the s sublevel; the second principal energy level has two sublevels (2s, 2p); the third, three (3s, 3p, 3d); and so on. This same information may be expressed by assigning each sublevel a quantum number ℓ and stating that ℓ can take on any integral value from 0 to (n − 1);

(3) *orbital* (Within a sublevel of quantum number ℓ, there are $2\ell + 1$ orbitals. An s sublevel ($\ell = 0$) has only one orbital; a p sublevel ($\ell = 1$) has three; a d sublevel ($\ell = 2$) five; and an f sublevel ($\ell = 3$) seven. Orbitals are assigned quantum numbers, m_ℓ, which take on all integral values from $-\ell$ to ℓ);

(4) *spin,* which is restricted to two possible values. These may be indicated as "up" and "down" (↑↓) or by assigning a quantum number m_s of $+\frac{1}{2}$ or $-\frac{1}{2}$. Experimentally, we find that no two electrons in an atom can have the same set of four quantum numbers (Pauli exclusion principle); it follows (why?) that a given orbital can contain no more than two electrons.

For many purposes, it is sufficient to describe the electronic structure of an atom by quoting its *electron configuration,* which tells us how many electrons are located in each sublevel. Thus, the electronic configuration for nitrogen (at. no. = 7), $1s^2 2s^2 2p^3$, tells us that there are 2 electrons in the 1s sublevel, 2 in the 2s, and 3 in the 2p. Electron configurations can be deduced by knowing the order in which sublevels are filled (p. 143, text) and recalling the total capacity of each sublevel (s = 2 × 1 = 2; p = 2 × 3 = 6; d = 2 × 5 = 10; f = 2 × 7 = 14). Examination of Table 6.4 shows us that this procedure is not infallible. Electrons occasionally show up in unexpected places (look at Cr and Cu, for example), but this need not concern us now.

Sometimes, we need to go one step further and give the *orbital diagram* of an atom, which indicates the number of electrons in each orbital and their relative spins. To derive orbital diagrams from electron configurations, we need only take account of two factors: (1) two electrons in the same orbital have opposed spins; (2) whenever possible, orbitals within the same sublevel will be half filled with electrons, all of which will have the same spin (Hund's rule). Thus, the orbital diagram for nitrogen is

1s	2s	2p
(↑↓)	(↑↓)	(↑) (↑) (↑)

The electron configurations of atoms correlate very simply with the positions of the corresponding elements in the periodic table. Elements in the same group of the table have the same number of electrons in the outermost principal energy level. For the A group elements, this number is equal to the group number. All the 1A elements have one "valence electron," all the 2A elements two, and so on. Since it is precisely these electrons that are involved in chemical bonding, it is hardly surprising that elements within a given group show similar chemical and physical properties.

The periodic table is a valuable device for predicting trends in properties of elements. We find, for example, that as one moves from left to right in the table, atomic radius and metallic character both decrease while ionization energy increases; as one moves down in the table, atomic radius and metallic character increase while ionization energy decreases. All of these trends, and many others, can be explained in terms of electronic structure; they are perhaps best recalled to mind by noting where elements fall in the table.

There are many questions we have yet to consider in discussing our modern model of atomic structure. Among them are: Why do ions form? (We need to further interpret ionization potentials.) Why do atoms bond together? Much of what composes the remainder of the text is an extension of atomic-molecular theory.

BASIC SKILLS

1. Use the Einstein equation (Equation 6.2) to relate the wavelength of a spectral line to the difference in energy between two levels.

A typical calculation of this type is shown in Example 6.1. See Problems 6.5 and 6.25. Note that energies are sometimes required (or given) in units other than ergs/atom. The following conversion factors may be helpful.

$$1 \text{ eV/atom} = 1.602 \times 10^{-12} \text{ erg/atom}$$

$$1 \text{ kcal/mole} = 6.949 \times 10^{-14} \text{ erg/atom}$$

2. Given the atomic number of an element, write the electron configuration of its atoms.

The electron configuration of an atom gives the number of electrons in each sublevel. In order to write an electron configuration you must know:

    a) the capacity of each sublevel (s = 2; p = 6; d = 10; f = 14).
    b) the order in which various sublevels are filled in atoms (1s, 2s, 2p, 3s, 3p, 4s, 3d, . . .). The complete list is given on p. 143 of the text.

- -

    What is the electron configuration of an atom of sulfur (atomic number = 16)? _____
    Using the rules given in (a) and (b) above, we fill each sublevel in succession until all 16 electrons are accounted for. First, we put 2 electrons in the 1s sublevel, then 2 in the 2s, 6 in the 2p, and 2 in the 3s sublevel. Stopping to count at this point, we find that we have used up: $2 + 2 + 6 + 2 = 12$ electrons. There are $16 - 12 = 4$ electrons left. The next sublevel to be filled is the 3p; since it has a capacity of 6 electrons, all 4 will fit nicely into it. The electron configuration is written: $1s^2 2s^2 2p^6 3s^2 3p^4$. Note that the number of electrons in each sublevel is indicated by a superscript at the upper right. Sometimes, electron configurations are condensed to show only those electrons beyond the preceding noble gas. Thus we might write for the sulfur atom:

$$[Ne]\ 3s^2 3p^4$$

where the symbol [Ne] represents the first 10 electrons.

- -

    Problems 6.8 and 6.28 are analogous to the example just worked. Problems 6.9 and 6.29 test your understanding of principles (a) and (b) referred to above.

    3. Given, or having derived, the electron configuration of an atom, draw its orbital diagram.
    To make this conversion, you must realize that:

    a) each sublevel is divided into orbitals capable of holding two electrons apiece. An s sublevel has 1 orbital, a p sublevel 3 orbitals, a d sublevel 5, and an f sublevel 7.
    b) When there are two electrons in an orbital, they have opposed spins, indicated by (↑↓).
    c) In a partially filled sublevel, there are as many half-filled orbitals as possible. Electrons in these orbitals have the same spins, e.g., (↑) (↑) (↑).

What is the orbital diagram for an atom of sulfur (atomic number = 16)?

Referring back to the electron configuration of sulfur, we might start by drawing the orbitals available in each of the sublevels involved. Applying rule (a),

1s	2s	2p	3s	3p
( )	( )	( ) ( ) ( )	( )	( ) ( ) ( )

Since the sublevels 1s through 3s are filled, we put two electrons with opposed spins in each of these sublevels to obtain

1s	2s	2p	3s	3p
(↑↓)	(↑↓)	(↑↓) (↑↓) (↑↓)	(↑↓)	( ) ( ) ( )

We must now distribute the remaining four electrons among the three 3p orbitals to leave as many half-filled orbitals as possible. A general approach here is to put single electrons with the same spin into successive orbitals until they are all half-filled. In this case, this consumes three of the four electrons we have: (↑) (↑) (↑). The remaining electron is added to the first orbital and given an opposed spin: (↑↓) (↑) (↑). The final orbital diagram is:

1s	2s	2p	3s	3p
(↑↓)	(↑↓)	(↑↓) (↑↓) (↑↓)	(↑↓)	(↑↓) (↑) (↑)

Problems 6.10 and 6.30 are analogous to this example. The reverse process, converting an orbital diagram to an electron configuration, is illustrated by Problems 6.11 and 6.31.

4. Give the four quantum numbers corresponding to various electrons in an atom.

The rules for assigning quantum numbers may be summarized as follows:

a) n = number of principal energy level = 1, 2, 3, 4, . . .

b) ℓ = 0, 1, . . . (n − 1). For an s electron, ℓ = 0; for a p electron, ℓ = 1; for a d electron, ℓ = 2; for an f electron, ℓ = 3.

c) m_ℓ can take on any integral value, including zero, ranging from $+\ell$ to $-\ell$.

d) m_s can be either + 1/2 or −1/2.

What are the quantum numbers of the four electrons in the 3p level of sulfur? _____

We might start by writing the six sets of quantum numbers that a 3p electron can have. We note from (a) above that n must equal 3 in each case. Rule (b) tells us that $\ell = 1$, since we are dealing with p electrons. The quantum number m_ℓ, which designates the orbital, may be +1, 0 or –1 (rule (c)). Finally, within a given orbital, m_s can be either +1/2 or –1/2. Hence, we have:

n	3	3	3	3	3	3
ℓ	1	1	1	1	1	1
m_ℓ	1	1	0	0	–1	–1
m_s	$\frac{+1}{2}$	$\frac{-1}{2}$	$\frac{+1}{2}$	$\frac{-1}{2}$	$\frac{+1}{2}$	$\frac{-1}{2}$

From the orbital diagram of sulfur derived on p. 78, we see that one 3p orbital is completely filled while the other two are half-filled. Arbitrarily choosing the orbital m_ℓ = +1 to be the one that is filled, we arrive at a possible set of four quantum numbers.

n	3	3	3	3
ℓ	1	1	1	1
m_ℓ	+1	+1	0	–1
m_s	$\frac{+1}{2}$	$\frac{-1}{2}$	$\frac{+1}{2}$	$\frac{+1}{2}$

This is not the only possible set; the following would be equally acceptable:

$$3,\ 1,\ +1,\ \frac{+1}{2};\ 3,\ 1,\ +1,\ \frac{-1}{2};\ 3,\ 1,\ 0,\ \frac{-1}{2};\ 3,\ 1,\ -1,\ \frac{-1}{2}$$

or:

$$3,\ 1,\ +1,\ \frac{+1}{2};\ 3,\ 1,\ 0,\ \frac{+1}{2};\ 3,\ 1,\ -1,\ \frac{+1}{2};\ 3,\ 1,\ -1,\ \frac{-1}{2}$$

However, the following sets would *not* be correct, the first because it implies that two orbitals (m_ℓ = +1 and m_ℓ = 0) are completely filled:

$$3,\ 1,\ +1,\ \frac{+1}{2};\ 3,\ 1,\ +1,\ \frac{-1}{2};\ 3,\ 1,\ 0,\ \frac{+1}{2};\ 3,\ 1,\ 0,\ \frac{-1}{2}$$

and the second because it shows the electrons in half-filled orbitals (m_ℓ = 0, m_ℓ = –1) to have opposed rather than parallel spins:

$$3,\ 1,\ +1,\ \frac{+1}{2}\ ;\ 3,\ 1,\ +1,\ \frac{-1}{2}\ ;\ 3,\ 1,\ 0,\ \frac{+1}{2}\ ;\ 3,\ 1,\ -1,\ \frac{-1}{2}$$

These principles are further illustrated in Problems 6.12 and 6.32.

5. Use the Periodic Table to predict:

   a) the physical properties of elements, given those of neighboring elements.

   The normal boiling points of Cl_2 and I_2 are $-35°C$ and $184°C$ respectively. Estimate the normal boiling point of Br_2. _____
   Since bromine falls between chlorine and iodine in the Periodic Table, we might expect the boiling point of Br_2 to be midway between those of Cl_2 and I_2.

$$\text{predicted bp} = \frac{-35°C + 184°C}{2} = 75°C$$

The observed boiling point is $59°C$, somewhat lower than predicted.

See also Example 6.2 and Problems 6.13 and 6.33.

   b) the relative sizes of different atoms.

   Of the four metals Na, Mg, K, and Ca, which would have the largest atomic radius? _____ the smallest? _____
   Since atomic radius increases as one moves down in the Table and decreases as one moves across from left to right, the atom closest to the lower left hand corner of the Table should be the largest. Locating these four elements in the Table, it is clear that the potassium atom should be the largest. By the same line of reasoning, the magnesium atom should be the smallest. The actual radii are: K = 2.31 Å, Ca = 1.97 Å, Na = 1.86 Å, Mg = 1.60 Å, consistent with our predictions.

See Problems 6.16 and 6.36. Note also that certain other properties of elements are correlated with atomic size. There is a direct correlation with metallic character (the larger the atom, the more metallic the element) and an inverse correlation with ionization energy (the larger the atom, the smaller the energy required to remove an electron).

   c) the formulas of binary or ternary compounds, given those of analogous compounds in the same groups of the Table.

The formula of potassium perchlorate is $KClO_4$. What is the formula of sodium periodate?

Since K and Na are in the same group (1A) and so are Cl and I (7A), we predict the formula to be $NaIO_4$. Note from the discussion on pp. 155-6 of the text that predictions of this type are not always reliable. Thus the reasoning here might lead us to predict the existence of lithium perfluorate, $LiFO_4$. In fact, no such compound has ever been isolated and there is good reason to believe that it never will be.

See Problems 6.14 and 6.34.

Problems

1. What is the energy difference in ergs corresponding to a spectral line at 3100 Å? What is the wavelength in Å of the line observed when an electron moves between two energy levels separated by 2.40×10^{-14} ergs?

2. Give the electron configurations of atoms of the elements $_9$F, $_{15}$P and $_{27}$Co.

3. Write orbital diagrams for each of the atoms in Problem 2.

4. Write a reasonable set of four quantum numbers for each of the electrons in the fluorine atom; do the same for each of the 3p electrons in phosphorus.

5. a) Given that the densities of Be, Ca, Na, and Al are 1.85, 1.55, 0.98, and 2.70 g/cm^3 respectively, predict the density of Mg (the observed value is 1.74 g/cm^3).
   b) Select the largest atom in each of the following sets: Ca, Mn, Br; Ca, Sr, Ba; Ca, Sc, Sr.
   c) Predict the formula of gallium selenate, given that of aluminum sulfate, $Al_2(SO_4)_3$.

6. Consider the unstable N^- ion.

   a) How many electrons are there in this ion?
   b) Write its electron configuration.
   c) Draw its orbital diagram.
   d) Give a reasonable set of four quantum numbers for each electron.

7. Which of the following electron configurations are reasonable for atoms in their ground (lowest energy) state? Which could represent excited states (one or more electrons promoted to a higher energy)? Which are impossible?

a) $1s^2 2s^2 2p^4$ b) $1s^2 2s^1 2p^5$ c) $1s^2 2p^6$ d) $1s^1 2p^7$
e) $1s^2 2s^3 2p^3$

8. Consider the ground state of the nitrogen atom, in which there are three 2p electrons. Which of the following represent reasonable sets of quantum numbers for these three electrons?

a) 2, 1, +1, +1/2   b) 2, 1, +1, +1/2
    2, 1, +1, –1/2   2, 1, 0, +1/2
    2, 1, 0, +1/2   2, 1, –1, +1/2

c) 2, 1, +1, –1/2   d) 2, 1, +1, +1/2
    2, 1, 0, –1/2   2, 1, 0, +1/2
    2, 1, –1, –1/2   2, 1, –1, –1/2

9. Consider the element Mn, atomic number 25.

a) Write its electron configuration.
b) Draw its orbital diagram. How many unpaired electrons does it have?
c) How would you expect its atomic radius to compare with that of Sc? La? Cs? Cl?
d) Which of the elements listed in (c) would you expect to be more metallic than Mn? Which would you expect to have a larger ionization potential?
e) Given that the formula of sodium permanganate is $NaMnO_4$, predict the formula of the corresponding compound containing the elements Cs, Tc and S.

10. According to the Bohr theory, the first two energy levels in the H atom have energies of -2.179×10^{-11} and -0.545×10^{-11} ergs.

a) For a given atom, what is the difference in energy in ergs between these two levels?
b) Express this energy difference in kcal/mole.
c) What is the wavelength in Å of the spectral line observed when an electron moves from one of these levels to the other?

– – – – – – – – – – –

11. Use the Bohr theory to calculate the electronic energies of the hydrogen atom for n = 1 to n = 6. Use these values to determine the transition responsible for the spectral line with wavelength of 1026 Å. What part of the electromagnetic spectrum includes this wavelength?

12. Calculate the value of the wavelength of light just sufficiently energetic to ionize a hydrogen atom. The ionization energy of hydrogen is 313.6 kcal/mole.

SELF-TEST

True or False

1. The volume of an atom is essentially that volume occupied ( ) by the electrons.

2. The greater the difference in energy between two levels, the ( ) longer the wavelength of the light emitted when an electron moves between them.

3. The atomic number is always equal to the number of ( ) electrons in a particular atom.

4. The radius of a negatively charged monatomic ion is always ( ) larger than the radius of the parent neutral atom.

5. The energy associated with an electron in a given atom is ( ) almost fully described by specifying its value of the quantum number n.

6. An orbital diagram is a geometrical representation of the ( ) shape of an orbital.

7. Without quantum theory, there probably would be no way of ( ) predicting the properties of element 110, yet to be discovered.

8. The energy associated with electromagnetic radiation ( ) increases in the order: x-ray, ultraviolet, visible, infrared.

Multiple Choice

9. What kind of attractive force seems to hold together the ( ) components of an individual atom?
  (a) gravitational   (b) magnetic
  (c) electrical   (d) the chemical bond

10. Experimental support for the arrangement of electrons in ( )
distinct energy levels is based primarily upon
    (a) the law of definite (constant) proportions
    (b) the law of conservation of energy
    (c) x-ray spectra of atoms
    (d) spectra from gas discharge

11. What is the total electron capacity of the energy level for ( )
which n = 4?
    (a) 8   (b) 16   (c) 18   (d) 32

12. The possible values of the magnetic quantum number m_ϱ of a ( )
3p electron are:
    (a) 0, 1, 2   (b) 1, 2, 3   (c) 1, 0, –1   (d) 2, 1, 0, –1, –2

13. The element whose neutral, isolated atoms have three half- ( )
filled 2p orbitals is:
    (a) $_5$B   (b) $_6$C   (c) $_7$N   (d) $_8$O

14. Which one of the following species has the same electron ( )
configuration as (is isoelectronic with) the argon atom?
    (a) Ne   (b) Na^+ (c) S^- (d) Cl^-

15. The electronic configuration of the oxide ion, O^{2-}, may be ( )
represented as:
    (a) $1s^2\,2s^2\,2p^4$ (b) $1s^2\,2s^2\,2p^2\,3s^2\,3p^2$ (c) $1s^2\,2s^2\,2p^6$ (d) $:O:^{2-}$

16. The formation of a gaseous +2 ion is probably easiest for ( )
    (a) $_1^2$H   (b) $_{12}$Mg   (c) $_{16}$S   (d) $_{20}$Ca

17. If ions were formed in the combination of tin ($_{50}$Sn) and ( )
another element, what would be the likely charge on the tin atoms?
    (a) –2   (b) +1   (c) +3   (d) +4

18. Which one of the following species would have an odd ( )
number of electrons?
    (a) N   (b) N^+ (c) NO_2^+ (d) electrons are always paired

19. Of the outermost electrons of the "inert" gases, it is true that: ( )
    (a) all are in filled p orbitals
    (b) all are paired
    (c) they complete a principal energy level
    (d) all are very difficult to remove from the atom

20. Which of the following species would you expect to have the ( )
largest radius?
    (a) $_{11}Na^+$ (b) $_{10}$Ne   (c) $_9F^-$ (d) $_8$O

21. If the following solids consisted of the most compact ( )
arrangement of atoms, which one would occupy the largest volume
per mole?

    (a) $_{19}$ K   (b) $_{37}$ Rb   (c) $_{38}$ Sr   (d) $_{48}$ Cd

22. The element with the largest value for the first ionization ( )
energy is located in the _____ portion of the modern extended
form of the periodic table.

    (a) upper right   (b) lower right
    (c) middle   (d) lower left

23. Which one of the following properties of an element would ( )
you expect *not* to be a periodic function of atomic number?

    (a) atomic volume   (b) specific heat
    (c) ionization potential   (d) boiling point

24. To estimate the density of hafnium, $_{72}$Hf, one would expect ( )
to average the densities of

    (a) La and Ta (b) Zr and Ce (c) Zr and U (d) Ta and Lu

25. The atomic number of the next-to-be-discovered noble gas, ( )
below radon, Rn, would be:

    (a) 109   (b) 118   (c) 173   (d) 222

26. The electronic configuration of the Fe atom is [Ar] $4s^2 3d^6$. ( )
The number of unpaired electrons in the orbital diagram is:

    (a) 0   (b) 2   (c) 4   (d) 6

27. Which of the following represents a reasonable set of ( )
quantum numbers for a 3d electron?

    (a) 3, 2, 1, 1/2   (b) 3, 2, 0, –1/2
    (c) neither of these   (d) both of these

28. Which of the following elements do you expect $_{30}$Zn to ( )
resemble most closely in its chemical properties?

    (a) $_{20}$Ca   (b) $_{21}$Sc   (c) $_{31}$Ga   (d) $_{48}$Cd

29. Which of the following could not be an orbital diagram for ( )
an atom in its ground state?

    (a) (↑↓) (↑↓) (↑↓)(↑)(↑) (b) (↑↓) (↑↓) (↑↓)(↑↓)(↑↓) (↑)
       1s 2s   2p   1s 2s   2p   3s
    (c) (↑↓) (↑↓) (↑↓)(↑↓)( ) (d) (↑↓) (↑↓) (↑↓)(↑↓)(↑↓) (↑↓)

30. The removal of one electron from a neutral atom requires   ( )
about the same energy for $_3$ Li, $_4$ Be, and $_5$ B: all their first ionization
energies are about 10^2 kcal/mol. The removal of a second electron,
however, is expected to be much more difficult for
(a) Li
(b) Be
(c) B
(d) no more for one than the others

31. Consider the schematic
energy level diagram shown for
the hydrogen atom. Which one of
the following interpretations is
not correct?

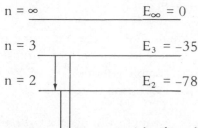

(a) The ground state and
the first excited state differ in
energy by (314–78) kcal/mol.
(b) An electron in level E_1 is closer to the nucleus, on the
average, than an electron in level E_2.
(c) Ionization is expected to occur with the absorption of at
least 314 kcal/mol.
(d) If an electron drops from level E_3 to level E_1, photons
having energies E_3 and E_1 are emitted.

SELF-TEST ANSWERS

1. **T**
2. **F**
3. **F** (Consider any monatomic ion.)
4. **T**
5. **T**
6. **F**
7. **F** (Consider that the "periodic law," and predictions based on it,
has a hundred year history.)
8. **F** (Energy and frequency decrease; wavelength increases.)
9. **c** (Nucleus-electron interactions involve opposite charges.)
10. **d**
11. **d** (Filled, it would be: $4s^2 4p^6 4d^{10} 4f^{14}$ Capacity $= 2n^2$.)
12. **c** (Corresponding to the three orientations of the p orbitals: p_x,
p_y, and p_z.)
13. **c** (With configuration: $1s^2 2s^2 2p^3$.)
14. **d** (Eighteen electrons.)

15. c   (Total of 10 electrons, filling lowest available levels.)

16. d   (Two valence electrons farthest removed from nucleus.)

17. d   (The first four ionization energies would all be relatively small. The four outer-level electrons, analogous to the four in carbon, should be relatively easy to remove.)

18. a   (Seven electrons in neutral atom = atomic number. $NO_2{}^+$ would have a total of $7 + 2(8) - 1 = 22$.)

19. b   (With regard to choice a, consider helium. Note that you expect the first ionization energy for Rn to be rather low.)

20. c   (Smallest nuclear charge exerts least attraction for the same number of electrons — for choices a-c; oxygen has fewer electrons.)

21. b

22. a

23. b   (That of the metals is, more or less, uniformly different from that of the nonmetals. Recall Law of Dulong and Petit.)

24. d   (Elements 71 and 73.)

25. b   (Element 104 would lie below $_{72}Hf$ . . .)

26. c   (Draw it.)

27. d

28. d

29. c

30. a   (Second electron comes from lower principle energy level.)

31. d   (Photon energy would be ΔE.)

SELECTED READINGS

Drago, R. S., *Qualitative Concepts from Quantum Chemistry*, Tarrytown-on-Hudson, N. Y., Bogden and Quigley, 1971.
A rather quantitative treatment of atomic and molecular structure; includes exercises with answers. Goes well beyond the text in treating the quantum mechanical model of atoms and bonding.

Hochstrasser, R. M., *Behavior of Electrons in Atoms: Structure, Spectra, and Photochemistry of Atoms*, New York, W. A. Benjamin, 1964.
The experimental background of modern atomic structure theory is the subject of this well-written book. About the same level as the text.

Lagowski, J. J., *The Structure of Atoms*, Boston, Houghton Mifflin, 1964.
Particularly useful are the discussions of the Bohr atom and of atomic spectra; with quotes from the pioneers.

Lewis, G. N., *Valence and the Structure of Atoms and Molecules*, New York, Dover, 1966.
A reprint of the 1923 classic, this presents a lucid and historical perspective on theory and experiment up to the early 1920's. It is to Lewis that we credit the idea of a bond as being a pair of electrons; his, too, the idea of the "electron dot diagrams."

Pimentel, G. C., and R. D. Spratley, *Chemical Bonding Clarified through Quantum Mechanics*, San Francisco, Holden-Day, 1969.

> *The first two chapters are easy reading and an interesting (often amusing) introduction to the modern theory of the electronic structure of atoms.*

Sanderson, R.T., Ionization Energy and Atomic Structure, *Chemistry* (May 1973), pp. 12-15.

> *A more detailed discussion of the experimental support for electron configurations.*

Seaborg, G. T., Prospects for Further Considerable Extension of the Periodic Table, *J. Chem. Ed.* (October 1969), pp. 626-634.

> *Predictions are made concerning the preparation and properties of elements not yet known, by one who has discovered his share of the elements.*

7

Chemical Bonding

QUESTIONS TO GUIDE YOUR STUDY

1. Why do bonds often form between neutral atoms (as well as between ions)? Why don't all the atoms in the universe bond together in one super molecule?

2. Is there anything common to all "types" of bonds, whether ionic, covalent, metallic, etc.?

3. How do you account for the observed differences in the strengths of bonds?

4. How are the size and shape of a molecule related to the electronic structures of the component atoms? Is there a simple, reliable way of predicting the geometry?

5. Can you predict formulas of compounds on a theoretical basis; in particular, in terms of the electronic structures of the atoms involved?

6. How does bonding theory help us explain such properties of materials as color, electrical conductivity, and boiling point?

7. How would you experimentally determine bond energies and distances?

8. How would you account for the fact that an atom may bond to just one other atom, sometimes to two others, or three . . . (e.g., C in CO, CO_2, CH_4)? Is the number of bonds predictable?

9. You have "constructed" individual atoms by filling atomic orbitals. Can you construct a molecule in an analogous manner?

10. Can you now give a more detailed description as to what occurs at the atomic-molecular level when a chemical reaction takes place?

11.

12.

YOU WILL NEED TO KNOW

Concepts

1. How to write electronic structures for representative elements — neutral atoms and also monatomic ions (say, for elements of atomic number 1 to 30, and others by analogy) — Chapters 6, 7

2. Which means that you need to be able to use the periodic table as a correlating tool (Example: the "valence electron" structure for $_{83}Bi$ is analogous to that of Sb and the other elements in Group 5A of the table: Bi — $6s^2 6p^3$; Sb — $5s^2 5p^3$) — Chapter 6

3. The shapes and relative sizes of atomic orbitals — Chapter 6

Math

1. How to calculate heats of bond formation from other heats (enthalpies) of reaction, and vice versa — Chapter 4

2. A familiarity with the geometry (symmetry and angles) of several plane and solid figures, particularly the tetrahedron, would be helpful — See a math text or the Readings of Chapter 9 and the Preface. A molecular model set would also be very useful in visualizing the geometric consequences of bonding.

CHAPTER SUMMARY

Most familiar materials consist *not* of individual isolated atoms (the subject of the preceding chapter), but of aggregates of atoms and ions. The existence of molecular clusters of atoms as well as the existence of condensed states support the idea that attractive forces, or "bonds," exist between atoms, ions, and even molecules. It is convenient, though somewhat arbitrary, to distinguish two kinds of forces: the generally stronger chemical bonds (this chapter) and the usually weaker intermolecular forces — or, perhaps, "physical bonds" (next chapter).

All of these attractive forces are electrostatic in nature. Their magnitudes include the very weak attraction between neighboring atoms in liquid helium, where 0.02 kcal is enough energy to separate a mole of the "bonded" particles into a monatomic gas; the very strong bond between atoms in a molecule of carbon monoxide, with a heat of bond formation of –257 kcal/mol; and almost anything in between. (Further examples of the range of attractive forces: Na_2 (g) has a heat of bond formation of –18 kcal/mol; for a mole of HF(g), this ΔH is –134 kcal; for a hydrogen bond, ΔH is typically between 1 and 10 kcal/mol.)

Again, it is convenient and a little arbitrary to classify chemical bonds according to the extent to which the electron distribution within atoms is changed when the atoms are strongly attracted to one another. The bond types discussed in this chapter are:

1) *Ionic bonds* between positive and negative ions. Ions are formed by the transfer of electrons from an element of low electronegativity (a metal) to an element of high electronegativity (a nonmetal). When two elements differ in electronegativity by more than 1.7 units, we expect the bonding between them to be predominantly ionic. We can readily deduce the charges of ions having a noble gas structure (1A, +1; 2A, +2; Al and 3B, +3; 6A, -2; 7A, -1). The transition metals commonly form cations with charges of +2 (Zn^{2+}, Cu^{2+}, Ni^{2+}, Fe^{2+}, Co^{2+}), +3 (Cr^{3+}, Fe^{3+}, Co^{3+}), or less frequently +1 (Ag^+). Many of the most common anions are polyatomic (OH^-, NO_3^-, SO_4^{2-}, CO_3^{2-}, PO_4^{3-}); the NH_4^+ ion is one of the very few polyatomic cations. Formulas of ionic compounds are readily deduced by imposing the condition of electroneutrality (e.g., $Zn(OH)_2$, $Cr(NO_3)_3$, Ag_2SO_4).

2) *Covalent bonds,* which consist of a pair of electrons shared between two atoms. We can expect covalent bonds to be formed when a nonmetal combines with another element which differs from it in electronegativity by less than 1.7 units. The greater the difference in electronegativity, the more polar will be the bond. In H_2 and Cl_2 where $\Delta EN = 0$, the covalent bond is nonpolar; i.e., the electron pair is equally shared by the bonded atoms. In contrast, the bond in the HCl molecule is polar; the bonding electrons are displaced toward the more electronegative Cl atom. The displacement of electrons that characterizes a polar covalent bond strengthens and shortens it; the formation of double bonds (two pairs of electrons between bonded atoms) or triple bonds (three pairs of electrons) has the same effect.

3) *Metallic bonds,* found in elementary substances of low electronegativity (metals) and in certain alloys of these elements. Metal atoms have too few valence electrons to form covalent bonds with all of their neighbors. Instead, valence electrons are "pooled" to form a negatively charged "glue" that holds the metal cations together. This simple picture of metallic bonding, which implies high electron mobility, offers a reasonable qualitative explanation of many of the general properties of metals.

A major portion of this chapter is devoted to the structure of molecules and polyatomic ions, both of which are held together by covalent bonds. Here, we are primarily interested in two factors: the way in which the valence electrons are distributed and the angles between the covalent bonds. The electron distributions in a wide variety of polyatomic ions and molecules can be represented satisfactorily by Lewis structures; these in turn can be derived by following the simple rules listed on p. 176 of the text.

Once you have arrived at the Lewis structure for a covalently bonded species, it is relatively easy to predict the bond angles and hence the geometry of the molecule or polyatomic ion. To do this, we apply the simple principle that electron pairs surrounding an atom are oriented to be as far apart as possible. This implies, for example, that four pairs of electrons around an atom will be directed toward the corners of a regular tetrahedron. The principle is readily extended to molecules containing multiple bonds by noting that, so far as geometry is concerned, a multiple bond behaves as if it were a single electron pair.

Knowing the geometry of a molecule and the relative electronegativities of its atoms, we can decide whether it is polar or nonpolar. Molecules in which polar bonds are arranged unsymmetrically with respect to one another (e.g., H_2O, NH_3) are polar. Molecules in which all the bonds are nonpolar (e.g., H_2, Cl_2) or in which all the polar bonds "cancel" one another by their symmetrical arrangement (e.g., CO_2, CCl_4) are nonpolar.

The concepts that we have just reviewed are fundamental to an understanding of chemical bonding and molecular structure. There are certain other less fundamental ideas with which you should be familiar. These include the terms "resonance" and "hybridization" which were grafted on to the atomic orbital approach to chemical bonding to rationalize some of its shortcomings. (Resonance structures constitute, in essence, an admission that we cannot represent a particular species in terms of one simple Lewis-type picture: the species has only one structure, and the problem is in finding a satisfactory way of describing it!)

Finally, there is included in this chapter an introduction to a somewhat more fundamental approach to molecular structure, the molecular orbital method. If you go on to take further courses in chemistry, you will hear a great deal more about molecular orbitals; in this chapter, we have done little more than outline the basic approach and apply it to some very simple molecules.

BASIC SKILLS

1. Using a Periodic Table if necessary, derive the formulas of ionic compounds.

This skill is illustrated in Example 7.1. Note that the formula of the compound is readily derived by applying the principle of electroneutrality (see Self Test question 9, Chapter 2 of this guide), provided you know the charges of cation and anion. The charges of the monatomic ions formed by the A group elements are easily deduced from their positions in the Periodic Table (1A = +1, 2A = +2, 3A = +3, 6A = -2, 7A = -1). Some of the more common ions of the transition metals are listed in Table 7.1. Table 7.2 gives the names and formulas of several polyatomic ions.

Problems 7.3 and 7.23 are entirely analogous to Example 7.1. In Problems 7.4 and 7.24, an additional step is required; after you decide upon the formula of the ionic solid, you must balance the equation for its formation.

2. Draw Lewis structures for molecules and polyatomic ions.

This skill is illustrated in Example 7.2. Problems 7.8, 7.9, 7.28 and 7.29 give you practice in drawing Lewis structures. Problems 7.12 and 7.32 are slightly more subtle, but the same principle is involved.

We cannot emphasize too strongly that the ability to draw Lewis structures is fundamental to a great many other skills. You cannot predict the geometry of a molecule, decide whether it is polar or nonpolar, write resonance forms, indicate the type of hybrid bonds present, and so forth unless you know its Lewis structure.

Frequently, more than one plausible Lewis structure can be written for a molecule or polyatomic ion. Such structures ordinarily differ only in their skeletons, i.e., the order in which atoms are bonded to one another. This point is discussed in Example 7.2. In part (b), the correct structure of the sulfate ion is deduced by applying the rule that in oxyanions, all the oxygen atoms are ordinarily bonded to the central, nonmetal atom. Again, in part (c), we arrive at the correct structure for formaldehyde by taking into account the fact that carbon always forms four bonds. As you progress in your study of chemistry, you will gradually add to your store of rules of this type.

3. Given or having derived the Lewis structure of a molecule or polyatomic ion, deduce its geometry.

The reasoning here is shown in Example 7.3. There are only two principles involved:

a) The electron pairs around an atom tend to be as far apart as possible. Thus, two electron pairs are oriented at 180° to each other; three electron pairs are directed toward the corners of an equilateral triangle; four electron pairs are directed toward the corners of a regular tetrahedron.

b) So far as geometry is concerned, a multiple bond behaves as if it were a single bond. Thus, in SO_2 (Example 7.3), we ignore the "extra" electron pair in the double bond and pretend that there are only three pairs of electrons around sulfur to arrive at a 120° bond angle.

Problems 7.13, 7.14, 7.33, and 7.34 give you a chance to apply these principles. Notice that, as pointed out previously, you need to know the Lewis structure before you can predict the geometry.

One point here sometimes causes confusion. In describing molecular geometry, we indicate only the positions of the atoms; the location of unshared electron pairs is not included. Thus we would not ordinarily refer to the NH_3 molecule as a "tetrahedron." Instead, we would describe only the positions of the three H atoms around the nitrogen, perhaps by calling it a "tetrahedron with one corner missing," or, better, a "trigonal pyramid" (i.e., a pyramid with a triangular base). This practice is followed because from experiment we can deduce only the positions of the atoms in the molecule; unshared pairs of electrons cannot be located.

4. Given or having derived the geometry of a molecule, predict whether or not it will be polar.

The general principle here is a simple one; a molecule can be nonpolar only if:

a) all the bonds are nonpolar as in H_2 or P_4.
b) the various polar bonds "cancel" each other as would be the case in the symmetrical molecules:

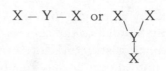

If the symmetry is destroyed by substituting an atom other than X as in:

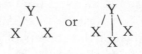

or by changing the bond angle as in:

the molecule is polar. This principle is illustrated in Example 7.4 and in Problems 7.15 and 7.35. Notice particularly in the example that you ordinarily need to know the geometry before you can predict polarity; geometry in turn can be deduced if the Lewis structure is known.

5. Given the Lewis structure of a molecule, give the hybridization around the central atom.

The hybrid orbitals discussed here include sp (two pairs of electrons), sp^2 (three pairs of electrons) and sp^3 (four pairs of electrons). Note that only one pair of electrons in a double bond can be hybridized; the other pair or pairs are not hybridized.

- -

What type of hybrid bond is formed by carbon in:

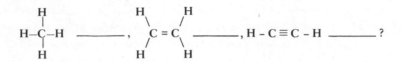

In CH_4 all four pairs of electrons around the carbon atom are hybridized; the four bonds are referred to as "sp^3 hybrids." In C_2H_4, only three pairs of electrons around an individual carbon atom are hybridized, and in C_2H_2 only two. The bonds are referred to as "sp^2 hybrids" and "sp hybrids" respectively.

- -

See Problems 7.17 and, particularly, 7.37.

*6. Given a Periodic Table, write the electron configuration of mon-atomic ions

For ions of elements in groups 1A, 2A, 3B, 6A and 7A this poses no problem, since they have the configuration of the nearest noble gas. For the ions of transition metals to the right of group 3B, you must realize that, unlike the situation in the corresponding atoms, d electrons are lower in energy than the s electrons in the next principal level (e.g., 3d is lower in energy than 4s).

- -

What is the electron configuration of K^+? _____ of Cr^{2+}? _____
The K^+ ion will have the electron configuration of the argon atom (Chapter 6): $1s^2 2s^2 2p^6 3s^2 3p^6$. In the Cr^{2+} ion, there are 22 electrons (two less than the 24 of the neutral Cr atom). The four electrons beyond the argon configuration will all be located in the 3d level: $1s^2 2s^2 2p^6 3s^2 3p^6 3d^4$ or [Ar] $3d^4$.

- -

See Problems 7.5 and 7.26. Note that you should also be able to write orbital diagrams for ions, using the principles discussed for atoms in Chapter 6.

*7. Given a Periodic Table, compare the relative sizes of ions and atoms.

The principles involved here are summarized on p. 167 of the text.

- -

Which one of the following species is the largest: Na, Na^+, Mg? _____ Cl^-, K^+, Ca^{2+}? _____

In the first case, the Na atom must be larger than the Na^+ ion (rule 2, p. 167). It must also be larger than the Mg atom (recall the discussion of relative sizes of atoms, Chapter 6).

The three species Cl^-, K^+, and Ca^{2+} all have the same number of electrons, 18, and hence the same electron configuration, that of the noble gas argon. They differ only in the number of protons in the nucleus (17 for Cl^-, 19 for K^+, 20 for Ca^{2+}). The ion with the smallest nuclear charge, Cl^-, will be the largest (rule 1, p. 167).

- -

8. Given a table of electronegativities (or at least a Periodic Table), compare the relative polarity of different bonds.

- -

Arrange the following bonds in order of increasing polarity: O—O, O—Cl, O—P _____

Using the table of electronegativities, we see that the differences in electronegativity are respectively 0, 0.5, and 1.4. Hence the bonds must become more polar in the order listed.

In this case, you could probably come to the same conclusion using only a Periodic Table. Clearly, the O—O bond is nonpolar since the atoms joined are identical. Recalling that electronegativity decreases as one moves down or to the left in the Table, we would expect phosphorus to be much less electronegative than oxygen. Conversely, chlorine lies below but to the right of oxygen in the Table, so these two elements should be quite similar in electronegativity. Thus we deduce that the order of increasing polarity is most likely: O—O < O—Cl < O—P.

- -

See Problems 7.6 and 7.26, both of which require that you use a table of electronegativities.

9. Given or having derived the Lewis structure of a molecule or polyatomic ion, draw possible resonance forms.

The phenomenon of resonance is discussed on pp. 178-9. Note that only two types of resonance are covered in this chapter. They involve moving either a multiple bond (SO_2, NO_3^-) or a single electron (NO) to a different position in a Lewis structure. Note further that one cannot get resonance forms by moving atoms around. These two observations may be helpful in working Problems 7.11 and 7.31.

10. Write molecular orbital diagrams for simple diatomic species.

The principles here are entirely analogous to those used in Chapter 6 to write atomic orbital diagrams for isolated atoms. Each MO, like each AO, can hold two electrons. When two orbitals of equal energy are available, electrons tend to enter singly with parallel spins (note, for example, the structures of B_2 and O_2 in Figure 7.16). Unfortunately, to work out MO diagrams, you must learn a new set of symbols (e.g., σ_{2s}^b, π_{2p}^*, ..) and a new filling order. The latter can be deduced from Figure 7.15 or 7.16. (Before you go to all this trouble, it would be a good idea to check with your instructor as to how much emphasis he wishes to place on this topic; opinions vary widely as to the importance of MO theory in general chemistry.)

Example 7.5 shows how an MO diagram can be derived for a simple diatomic species. Problems 7.18, 7.19, 7.38, and 7.39 further test this skill and illustrate some of the applications of MO structures.

Problems

1. Give the formulas of the following ionic compounds.

a) lithium bromide
b) lithium oxide
c) lithium sulfate
d) strontium sulfide
e) strontium iodide

f) strontium nitrate
g) aluminum phosphate
h) aluminum oxide
i) aluminum nitrate

2. Draw Lewis structures for

a) FCl   b) BrO$^-$ c) Cl$_2$O   d) NO$_2$$^-$ e) SeO$_3$

3. Predict the geometry of the species in (2).

4. Which of the molecules in (2) are polar?

5. Give the hybridization of N in NO$_2$$^-$; of O in Cl$_2$O.

*6. Write electron configurations for

a) H$^-$ b) Mg^{2+} c) Mn^{2+}

*7. Choose the smallest species in each set.

a) Li$^+$, H$^-$, Be^{2+} b) Co, Co^{2+}, Cr   c) F, O^{2-}, F$^-$

*8. Arrange the following bonds in order of increasing polarity, using only a Periodic Table: N–O, P–O, Ge–O, O–O

*9. Which of the species in (2) would you expect to show resonance?

*10. Write MO diagrams (consider only the 2nd principal energy level) for

    a) NO^- b) O_2^-

– – – – – – – – – – – –

11. For each of the following species: OH^-, CH_4O, SCN^-, ClO_3^-, SO_3

    a) Write a reasonable Lewis structure.
    b) Give each bond angle (e.g., 109°, 120°, ...).
    c) State whether the species is a dipole.

12. Consider the acetic acid molecule: $H_3C - C - O - H$
$$\overset{\|}{\underset{O}{}}$$

    a) Sketch the molecule, giving each bond angle.
    b) What is the hybridization around each of the carbon atoms?

13. For each of the following ionic compounds, give the simplest formula, the electron configuration of the positive ion, and the Lewis structure of the negative ion.

    a) sodium nitrate   c) calcium phosphate
    b) potassium carbonate   d) nickel(II) sulfate

SELF-TEST
True or False

    1. Chemical bonds form because one or more electrons are ( )
attracted simultaneously by two or more nuclei.

    2. Chemical bonds almost never form unless half-filled orbitals ( )
are available.

    3. A reasonable formula for a compound of aluminum and ( )
bromine would be Al_3Br.

    4. The strength of the attractive force between two bonded ( )
atoms increases, and the length of the bond increases, in the order:
single bond – double bond – triple bond.

    5. The coefficient of boron, $_5B$, in the equation for the reaction ( )
of boron with oxygen, can be predicted from the electronic
structures of boron and of oxygen.

6. A binary compound consisting of an element having a low ( )
ionization potential and a second element having a high electronegativity is likely to possess covalent bonds.

7. Experimental results support the idea that a certain molecule, ( )
AB_2, is linear (that is, all three nuclei lie along a straight line). This
must mean that there are no unshared electrons about the central
atom.

8. There are 24 valence electrons in the $SO_3{}^{2-}$ ion. ( )

Multiple Choice

9. The ion, Ni^{2+}, would have the electron configuration: ( )
   (a) [Ar] $3d^8 4s^2$　　　　　　　　(b) [Ar] $3d^8$
   (c) [Ar] $3d^7 4s$　　　　　　　　(d) [Ar] $3d^6 4s^2$

10. Of the elements listed below, which would you expect to ( )
have the largest attraction for bonding electrons?
   (a) $_3$Li　　　　(b) $_{13}$Al　　　　(c) $_{26}$Fe　　　　(d) $_{16}$S

11. The forces most suited to account for the fact that atoms ( )
often combine to form molecules are
   (a) nuclear　　(b) magnetic　　(c) electrical　　(d) polar

12. Which one of the following contains both ionic and covalent ( )
bonds?
   (a) NaOH　　　(b) HOH　　　(c) $C_6 H_5 Cl$　　　(d) SiO_2

13. For the reactions: ( )

$$C(g) + 4 H(g) \rightarrow CH_4 (g) \qquad \Delta H = -395 \text{ kcal/mole}$$
$$2 C(g) + 4 H(g) \rightarrow H_2 C = CH_2 (g) \qquad \Delta H = -541 \text{ kcal/mole}$$

The enthalpy of bond formation of the C=C bond, in kcal/mole, is: ( )
   (a) −99　　　(b) +146　　　(c) −170　　　(d) −146

14. Which one of the following species would have an unpaired ( )
electron?
   (a) SO_2　　　(b) $NO_2{}^-$　　　(c) $NO_2{}^+$　　　(d) NO_2

15. Which of the following bonds would be the least polar? ( )
   (a) H–F　　　(b) O–F　　　(c) Cl–F　　　(d) Ca–F

16. The fact that all bonds in SO_3 are the same is accounted for ( )
by
   (a) the idea of resonance
   (b) the Lewis octet rule
   (c) electron repulsion theory
   (d) sulfur always forms three bonds

17. Indicate which one of the following does not obey the octet rule: ( )

    (a) NO_3^- (b) O_3 (c) HCN   (d) NO

18. The molecular shape of the compound PH_3 is predicted as being ( )

    (a) planar   (b) linear   (c) pyramidal   (d) something else

19. Which species is most likely to be planar? ( )

    (a) NH_4^+ (b) CO_3^{2-} (c) SO_3^{2-} (d) ClO_3^-

20. The electron pairs on Cl can be considered as being approximately tetrahedrally directed in ( )

    (a) ClO^- (b) ClO_2^- (c) ClO_4^- (d) all of these

21. Indicate which one of the following is definitely polar: ( )

    (a) O_2 (b) CO_2 (c) BF_3 (d) C_2H_3F

22. The fact that the BeF_2 molecule is linear implies that the Be–F bonds involve ( )

    (a) sp hybrids   (b) sp^2 hybrids
    (c) dsp^2 hybrids   (d) resonance

23. sp^3 hybridization is most likely not involved in a description of ( )

    (a) H_3O^+ (b) CCl_4 (c) NH_4^+ (d) C_2H_2

24. Although it is known that the actual bonding sequence of atoms in the nitrous oxide molecule is N–N–O, an electronic structure based on the sequence N–O–N (i.e., oxygen bonded to two nitrogens): ( )

    (a) cannot be drawn
    (b) can also be drawn, and predicts a polar molecule
    (c) can also be drawn, and involves unpaired electrons
    (d) can also be drawn, and obeys the octet rule

25. The hybridization of boron in B_2H_4 is expected to be ( )

    (a) sp   (b) sp^2 (c) sp^3 (d) dsp^2

26. What type of hybrid orbitals are used by carbon in formaldehyde, CH_2O? ( )

    (a) sp   (b) sp^2 (c) sp^3 (d) d^2sp^3

27. Metallic substances are distinguished from other classes of chemical substances by their ( )

    (a) high boiling points
    (b) being solids at room temperature
    (c) ability to conduct electricity in the liquid state
    (d) high conductivity in the solid state

28. The description of the electronic structure of O_2 that best ( )
accounts for both the bond strength and the magnetic properties is
given by
    (a) :Ö=Ö:
    (b) :Ö–Ö:
    (c) a resonance hybrid of structures (a) and (b)
    (d) molecular orbital theory

29. Of the following interactions between two atoms or ( )
molecules or ions, the strongest is most likely to be
    (a) gravitational attraction
    (b) ionic bond
    (c) dipole-dipole forces between polar molecules
    (d) hydrogen bond

30. All chemical bonds are the result of ( )
    (a) the magnetic interaction of electrons
    (b) the interaction of nuclei
    (c) differences in electronegativity
    (d) the interaction of electrons and nuclei

31. The electrons generally involved in bonding ( )
    (a) are those that lie closest to the nucleus
    (b) are those for which the ionization energies are small
    (c) end up being transferred from one atom to another
    (d) occupy s atomic orbitals

32. The fact that ionization energies for "valence" electrons in ( )
units of kcal/mole (rather than the usual eV/atom) are of the same
order of magnitude as heats of reaction for chemical changes
    (a) is a mere coincidence involving units
    (b) supports the idea that only the outer-level electrons are
        generally involved in molecular rearrangements
    (c) supports the idea that chemical changes involve moles of
        atoms and not simply individual atoms
    (d) has no apparent rationale

33. Which one of the following would have a Lewis structure ( )
most like that of $CO_3{}^{2-}$?
    (a) CO_2 (b) $SO_3{}^{2-}$ (c) $NO_3{}^-$ (d) O_3

34. The weakest N to N bond among the following is most likely ( )
that in
    (a) :N≡N: (b) F–N̈=N̈–F
    (c) F–N̈–N̈–F   (d) H–N̈=N=N̈:
          | |
         F F

35. A bond angle of 120° appears in species with which ( )
geometry?
    (a) linear   (b) square
    (c) planar, triangular   (d) tetrahedral

36. Which of the following is a correct Lewis structure for ( )
C_2H_4O?

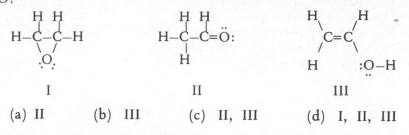

        I   II   III

    (a) II   (b) III   (c) II, III   (d) I, II, III

SELF-TEST ANSWERS

1. **T**
2. **F** (Look up "coordinate covalent bond". Note that hybridization often results in the unpairing of electrons. Consider: Be in BeF_2)
3. **F** ($AlBr_3$, where ions expected are Al^{3+} and Br^-.)
4. **F** (Bond length decreases.)
5. **T** (Consider atom ratio of B to O needed for octets; or, consider what ions are likely to form.)
6. **F** (Ionic: combination is of a metal and a nonmetal.)
7. **F** (Example of reasonable structure: $:\ddot{B}=A=\ddot{B}:$)
8. **F** (Charge contributes 2 electrons.)
9. **b** (Electrons are removed first from the highest principal energy level.)
10. **d**
11. **c**
12. **a** (Na⁺ is bonded ionically to O–H⁻, which itself is held together by a covalent bond.)
13. **d**
14. **d**
15. **b**
16. **a** (Try drawing *only one* Lewis structure!)
17. **d** (An odd number of electrons.)
18. **c**
19. **b** (Apply "electron pair repulsion" ideas.)
20. **d**
21. **d**
22. **a**
23. **d** (Rather, sp hybrids.)

24. d (For example: :N̈=O=N̈:)
25. b
26. b (Actual structure: H–C–H)

$$\underset{:\ddot{O}:}{\overset{\|}{}}$$

27. d
28. d
29. b (See Chapters 8 and 11 for further discussion of hydrogen bonding.)
30. d (See question 1.)
31. b (Outer-level or "valence" electrons.)
32. b (One can think of the formation of an ionic bond as involving ionization.)
33. c (They are isoelectronic.)
34. c
35. c
36. d (All obey the octet rule.)

SELECTED READINGS

Campbell, J. A., *Why Do Chemical Reactions Occur?*, Englewood Cliffs, N. J., Prentice-Hall, 1965.
  The discussion of bonding points out that sharp lines cannot be drawn between bond types. Observed heats of bond formation vary almost continuously, gradually, from one extreme to the opposite.)
Ferreira, R., Molecular Orbital Theory: An Introduction, *Chemistry* (June 1968), pp. 8-15.
  A graphic introduction that should be well within reach.
Lagowski, J. J., *The Chemical Bond*, Boston, Houghton Mifflin, 1966.
  A survey of bonding theories, with quotations from the original works. Slights MO theory.
Pauling, L., *The Nature of the Chemical Bond*, 3rd Ed., Ithaca, N. Y., Cornell, 1960.
  A very extensive discussion of valence bond theory and its applications — by its creator. Written with style (almost unique in modern scientific literature). This classic is mainly for the advanced student.
Selig, H. and others, The Chemistry of the Noble Gases, *Scientific American* (May 1964), pp. 66-77.
  Discusses in a conversational way the discovery of the compounds of the noble gases and their bonding. Also reviews bonding theory.
Ward, R., Would Mendeleev Have Predicted the Existence of XeF_4?, *J. Chem. Ed.* (May 1963), pp. 277-279.
  The author's answer is Yes, and more! An interesting, compact synthesis of the ideas found in Chapters 6 and 7 of the text.

Also see the readings of Chapter 6, by Pimentel and by Lewis.

Physical Properties of Molecular Substances; Nature of Organic Compounds

QUESTIONS TO GUIDE YOUR STUDY

1. How does modern chemical theory explain the properties of bulk samples of matter? What factors at the atomic-molecular level determine these properties?

2. What are the distinguishing properties of covalently bonded species? How are they explained? What kind of experimental observation would tell you a substance was covalently bonded?

3. What kind of experimental evidence would tell you that you were dealing with interatomic forces; with intermolecular forces? What are their relative magnitudes?

4. How would you know that you were working with very large, or very small, molecules?

5. What factors favor the formation of small, discrete molecules; of very large, extensive molecules? Is there a correlation with the Periodic Table?

6. What are polymers? What are their special properties and how are they explained? (Think of some common examples of polymeric materials.)

7. Can you now more fully explain, in terms of interatomic and intermolecular forces, what you can directly observe during a chemical reaction? What is happening among the molecules that results in the gross changes you detect?

8. How does the electronic nature of a bonded group of atoms, e.g., $-C-O-H$, determine the physical and chemical properties characteristic of that group?

105

9. How do you account for the vast number and variety of carbon compounds?

10. How is it possible for two different compounds to have the same molecular formula?

11.

12.

YOU WILL NEED TO KNOW

Concepts

1. How to write Lewis structures – Chapter 7

2. How to recognize, from a chemical formula, whether a species is likely to be molecular (i.e., has covalent or polar covalent bonds; a prediction based primarily on electronegativity differences) – Chapter 7

3. How to predict the geometry and polarity of molecules – Chapter 7

Math

Finally, a chapter without math!

CHAPTER SUMMARY

In this chapter, we continue the discussion of chemical bonding and extend the atomic-molecular model to include the weaker forces (between molecules) which account particularly for physical properties. Two additional topics are also related to bonding and molecular structure.

1) *Intermolecular attractive forces,* whose magnitudes determine properties of molecular substances – such properties as melting point, boiling point, vapor pressure, and heat of vaporization. We recognize three different types of molecular forces:

   a) *dipole forces,* which operate between polar molecules and account for the fact that these substances ordinarily have higher melting and boiling points than do nonpolar substances of comparable molecular weight.

   b) the *hydrogen bond,* a particularly strong type of dipole force, which arises between molecules in which hydrogen is bonded to a small, highly electronegative atom (O, N, F). These bonds are found in a few inorganic compounds (e.g., H_2O, NH_3, HF, HCN) and a wide variety of organic compounds, including alcohols, acids, and, perhaps most important of all, proteins.

c) *dispersion forces,* which operate between all molecules, polar or nonpolar. These forces, arising from the distortion of the electron density of a molecule by its neighbor, increase in magnitude with molecular size. This explains the general observation that melting point and boiling point increase with molecular weight.

2) The structure and properties of *macromolecular substances,* where large numbers of atoms are held together by covalent bonds to form a huge molecular aggregate. Many familiar substances fall in this category. They include the allotropic forms of carbon, silicon dioxide and related compounds (quartz, sand, glass, asbestos, and others) and such familiar organic polymers as polyethylene and nylon. We have attempted to show in this chapter how the two- or three-dimensional structure of these substances determines their properties and hence their commercial uses.

3) The structure and properties of *organic compounds,* to which a major portion of this chapter is devoted. The simplest organic compounds are the hydrocarbons, which contain only carbon and hydrogen. Among hydrocarbons, we distinguish between paraffins or alkanes, in which all the bonds between carbon atoms are single (e.g., CH_4, C_2H_6, C_3H_8,...C_nH_{2n+2}); alkenes, where there is one double bond (e.g., C_2H_4, C_3H_6,...C_nH_{2n}); and alkynes, with one triple bond (C_2H_2, C_3H_4,...C_nH_{2n-2}). Still another class of hydrocarbons is made up of the aromatics, which may be considered as derivatives of benzene, C_6H_6.

Many of the most familiar organic compounds contain the three elements carbon, hydrogen and oxygen. Such compounds are conveniently classified according to the functional groups present. All alcohols contain the group $-\overset{\displaystyle |}{\underset{\displaystyle |}{C}}-OH$, all organic acids the group $-\overset{\displaystyle ||}{\underset{\displaystyle O}{C}}-OH$, and so on (see Table 8.4). All compounds containing a given functional group show similar physical properties: within a series of alcohols, or acids, etc., there is a gradation in physical properties with increasing molecular weight. In this sense, there is a resemblance to groups of elements in the periodic table.

One reason for the multiplicity of organic compounds is the phenomenon of isomerism, which is very common in organic chemistry. We ordinarily find that for a given molecular formula there are several different compounds with quite different properties. Examples include n-butane and iso-butane (molecular formula C_4H_{10}); n-propyl alcohol and isopropyl alcohol (C_3H_8O); butyl alcohol and diethyl ether ($C_4H_{10}O$). Noting that each of these compounds contains at least two carbon atoms bonded to each other, we see one reason why isomerism is much more common in organic than in inorganic chemistry. Atoms of elements other than carbon very rarely bond to one another to give structures of the type –X–X– (Can you explain why?)

━━━━

BASIC SKILLS

1. Classify a substance as to the type of bonding present and make qualitative predictions concerning its physical properties.

The physical properties of four kinds of substances discussed in Chapters 7 and 8 are summarized in Table 1 on the next page. (Note that our attention in this chapter will be focused primarily on intermolecular forces, the forces between the structural units we call molecules – items 2 and 3 in the table.)

– –

Consider the five substances: CO_2, SiC, Mg, KBr, CCl_4. Which are ionic?_____ molecular?_____ macromolecular?_____ metallic?_____ Which would you expect to be solids at room temperature?_____ Which would you expect to boil below 500°C?_____ Which would conduct electricity in the (pure) liquid state? _____

Mg is clearly a metal (note its position in the Periodic Table). KBr would be expected to be ionic, as are most compounds derived from metals. CO_2 and CCl_4 are molecular, as are most compounds containing only nonmetals. SiC has a structure similar to carbon; i.e., it is macromolecular.

Once the substances have been classified as to type, Table 1 can be used to predict their properties. SiC (macromolecular), Mg (metallic), and KBr (ionic) are all solids at room temperature; CO_2 (molecular) is a gas and CCl_4 (molecular) is a liquid. Only CO_2 and CCl_4 boil below 500°C. KBr (molten) and Mg (molten or solid) are conductors.

– –

This skill may be applied in Problem 8.2c, d, and e, and Problems 8.22 b, c, and f.

2. Classify molecular substances as to the types of intermolecular forces present.

There are three types of intermolecular forces:

    a) dispersion forces, common to all molecular substances;
    b) dipole forces, restricted to polar molecules;
    c) hydrogen bonds, found in polar molecules where H is bonded to F, O, or N.

TABLE 1 – MOLECULAR STRUCTURE AND PROPERTIES OF BULK MATTER

Structural Units	Forces Within Units	Forces Between Units	Properties	Examples
1. Ions	Covalent bond in polyatomic ions (e.g., CO_3^{2-})	Ionic bond	Hard, brittle solid (high mpt, bpt); conducts in melt and water solution; soluble in polar solvent	$NaCl$, MgF_2, $CaCO_3$
2. Molecules (a) nonpolar	Covalent bond	Dispersion force	Soft solid, or liquid or gas (low mpt, bpt); nonconductor; soluble in nonpolar solvent	H_2, CH_4, BF_3
(b) polar	Covalent bond	Dispersion force and dipole force; hydrogen bond	Low mpt, bpt (but higher than for nonpolar); often conducts in water solution; more soluble in polar solvent than nonpolar	HCl, NH_3, CH_3OH
3. Macromolecule, network	Covalent bond; often, any other as well	Dispersion force, dipole force, hydrogen bond may be present	Hard solid (high mpt, bpt); may decompose on heating; usually nonconductor; insoluble	C, SiO_2, cellulose
4. Cations and mobile electrons	. . .	Metallic bond	Variable mpt, high bpt; variable hardness; conductor; insoluble in each other	Na, Fe, brass

Approximate order of bond strength: Dipole $\sim$ Dispersion $<$ Hydrogen $<<$ $\begin{cases} \text{Ionic} \\ \text{Covalent} \\ \text{Metallic} \end{cases}$

What types of intermolecular forces are present in H_2?_____
CCl_4?_____ $CHCl_3$?_____ H_2O?_____

The H_2 and CCl_4 molecules are nonpolar (Chapter 7). Hence they are attracted to each other only by dispersion forces. The $CHCl_3$, and H_2O molecules are both polar (Chapter 7). In $CHCl_3$, there are dipole as well as dispersion forces. In H_2O, hydrogen bonds as well as dispersion forces are present.

Note that in order to apply this skill, you must first decide whether the molecule is polar or nonpolar. See Problems 8.3, 8.4, 8.23, and 8.24.

3. Classify molecular substances as to their relative boiling points.

The principles involved here are:

a) among substances of similar structure, boiling point ordinarily increases with molecular weight;

b) among substances of comparable molecular weight, those which are more polar or less compact ordinarily boil higher;

c) hydrogen bonded species have unusually high boiling points.

Which would you expect to have the higher boiling point, H_2 or N_2?_____ N_2 or CO?_____ n–C_4H_{10} or iso–C_4H_{10}? _____

N_2 and H_2 are both nonpolar molecules; since N_2 has the higher molecular weight, we would expect it to have the higher boiling point ($-196°C$ vs. $-253°C$ for H_2). N_2 and CO have the same molecular weight, 28; since CO is polar, it would be expected to have the higher boiling point ($-192°C$ vs. $-196°C$ for N_2). Of the isomers of butane, the "long, thin" n-C_4H_{10} molecule is expected to boil higher than the more compact iso-C_4H_{10}.

$$H_3C - CH_2 - CH_2 - CH_3 \qquad\qquad H_3C - \overset{\displaystyle H}{\underset{\displaystyle CH_3}{C}} - CH_3$$

n–C_4H_{10} ; bpt = $0°C$ $\qquad\qquad$ iso–C_4H_{10} ; bpt = $-10°C$

See Problem 8.2a, b, f; Problem 8.22a, d, e; and Problem 8.8.

4. Given the molecular formula of a simple organic compound, write structural formulas for its different isomers.

Write structural formulas for the isomers of pentane, C_5H_{12}. _____

Here it is best to proceed systematically. We start by putting all five carbon atoms in a straight chain to obtain:

$$H_3C - CH_2 - CH_2 - CH_2 - CH_3 \qquad (1)$$

Next we consider a four-carbon chain with one branch. There is only one isomer of this type, which we might represent as:

$$
\begin{array}{c}
\text{H} \\
| \\
H_3C - \ \ C \ \ - CH_2 - CH_3 \\
| \\
CH_3
\end{array}
\qquad (2)
$$

Finally we go to a three-carbon chain with two branches. Again, there is only one isomer.

$$
\begin{array}{c}
CH_3 \\
| \\
H_3C - \ \ C \ \ - CH_3 \\
| \\
CH_3
\end{array}
\qquad (3)
$$

You may be tempted to write other isomers which may look different on paper. By using ball and stick models, you can convince yourself that any such structures are identical to one of those shown above.

Problem 8.5 is identical to the example just worked; Problem 8.25 extends it one more step. In Problems 8.7 and 8.27, you draw isomers for unsaturated hydrocarbons. Problem 8.10 is clearly the most difficult of this type, since you have several different molecular formulas to work with (C_3H_7Cl, $C_3H_6Cl_2$, ... C_3Cl_8).

5. Given or having derived the structural formula of a hydrocarbon, classify it as a paraffin, olefin, acetylene, or aromatic hydrocarbon.

If you are given directly the structural formula, this type of problem is easy. All you need to remember is that in a paraffin all the C–C bonds are single; in an olefin there is a double bond; in an acetylenic hydrocarbon, there is a triple bond. Finally, in an aromatic hydrocarbon there is a benzene ring.

Frequently, all you have to work with is a molecular formula (e.g., C_4H_8), as in Problems 8.14 and 8.34. Here, you must first derive the structural formula. Some hints: the general formula of a paraffin is C_nH_{2n+2}; an olefin has two fewer hydrogen atoms (the two carbon electrons forming a double bond instead), C_nH_{2n}; an acetylene, two fewer yet, $C_nH_{2\,n-2}$. Thus, the formulas C_4H_{10}, C_4H_8, and C_4H_6 represent a paraffin, an olefin, and an acetylene respectively.

6. Given or having derived the structural formula of an oxygen-containing organic compound, classify it as an alcohol, ether, aldehyde, ketone, acid, or ester.

These compounds are identified by the type of functional group present (Table 8.4). Applying the information in this table, you should have no difficulty with Problems 8.11 and 8.31. Problems 8.13 and 8.33 involve an extra step; here you first have to derive the structural formula from the molecular formula. Some points worth noting:

— oxygen-containing organic compounds, like hydrocarbons, can have isomers;
— every alcohol (except CH_3OH) is isomeric with at least one ether;
— every aldehyde (except HCHO and H_3C-CHO) is isomeric with at least one ketone;
— every acid (except H-COOH) is isomeric with at least one ester.

7. Given the structural formula of a monomer, write a portion of the structure of the corresponding polymer.

————————————————————————————————————

Sketch a portion of the addition polymer formed from vinyl chloride,

$$H_2C = \overset{\overset{\displaystyle H}{|}}{C} - Cl \;\underline{\hspace{1.5cm}}\; ;\; \text{the polymer formed by condensing}$$

$$HO - \overset{\overset{\displaystyle O}{\|}}{C} - CH_2 - \overset{\overset{\displaystyle O}{\|}}{C} - OH \;\text{with}\; \overset{\overset{\displaystyle H}{|}}{H}N - CH_2 - N\overset{\overset{\displaystyle H}{|}}{H} \underline{\hspace{1.5cm}} .$$

In the first case, we start by opening up the double bond to give a single,

unpaired electron on each carbon atom: $\cdot \overset{\overset{\displaystyle H}{|}}{\underset{\underset{\displaystyle H}{|}}{C}} - \overset{\overset{\displaystyle Cl}{|}}{\underset{\underset{\displaystyle H}{|}}{C}} \cdot$

Other molecules can now bond at each end of the monomer:

$$\cdot \overset{H}{\underset{H}{C}} - \overset{Cl}{\underset{H}{C}} \cdot + \cdot \overset{H}{\underset{H}{C}} - \overset{Cl}{\underset{H}{C}} \cdot + \cdot \overset{H}{\underset{H}{C}} - \overset{Cl}{\underset{H}{C}} \cdot \rightarrow \cdot \overset{H}{\underset{H}{C}} - \overset{Cl}{\underset{H}{C}} - \overset{H}{\underset{H}{C}} - \overset{Cl}{\underset{H}{C}} - \overset{H}{\underset{H}{C}} - \overset{Cl}{\underset{H}{C}} \cdot$$

Since there are still unpaired electrons at each end of the molecule, polymerization can continue to give a product of very high molecular weight.

In condensation polymerization, a small molecule such as H_2O is split out from between two different monomer molecules:

$$HO - \overset{O}{\underset{\|}{C}} - CH_2 - \overset{O}{\underset{\|}{C}} - OH + H - \overset{H}{\underset{|}{N}} - CH_2 - \overset{H}{\underset{|}{N}} - H + HO - \overset{O}{\underset{\|}{C}} - CH_2 - \overset{O}{\underset{\|}{C}} - OH \rightarrow$$

$$- \overset{O}{\underset{\|}{C}} - CH_2 - \overset{O}{\underset{\|}{C}} - \overset{H}{\underset{|}{N}} - CH_2 - \overset{H}{\underset{|}{N}} - \overset{O}{\underset{\|}{C}} - CH_2 - \overset{O}{\underset{\|}{C}} -$$

Problems

1. Classify each of the following substances as ionic, molecular, macromolecular, or metallic.

   a) Al   b) Br_2 c) CaF_2 d) NO_2 e) SiO_2

   Which three of these substances are solids at room temperature?
   Which one of these substances is an electrical conductor in the solid state?
   Which of these substances are electrical conductors in the liquid state?

2. List the type of intermolecular forces present in

   a) N_2 b) HF   c) NO   d) NH_3 e) CH_4

3. Which member of each of the following pairs would you expect to have the higher boiling point?

   a) ICl or Br_2 b) CH_4 or CCl_4
   c) H_2O or H_2S d) He or Ar

4. Write structural formulas for all the isomers of the olefin whose molecular formula is C_4H_8.

5. Classify each of the following as to the type of hydrocarbon.

   a) $H_3C - \overset{H}{\underset{CH_3}{\overset{|}{\underset{|}{C}}}} - CH_3$

   b) [benzene ring with H_3C and CH_3 substituents and CH_3]

   c) $H_3C - \overset{H}{\underset{|}{C}} = CH_2$

   d) $H - C \equiv C - CH_3$

   e) $\overset{H}{\underset{H_3C}{}} C = C \overset{CH_3}{\underset{H}{}}$

6. Classify as to type each of the following oxygen-containing organic compounds.

a) $H_3C - \underset{\underset{OH}{|}}{\overset{\overset{H}{|}}{C}} - CH_3$ b) $H_3C - \underset{\underset{O}{\|}}{C} - O - CH_3$

c) $H - \underset{\underset{O}{\|}}{C} - CH_2 - CH_3$ d) $H_3C - \underset{\underset{COOH}{|}}{\overset{\overset{H}{|}}{C}} - CH_3$

e) $H_3C - O - CH_2 - CH_3$ f) $H_3C - O - \underset{\underset{O}{\|}}{C} - CH_2 - CH_3$

7. Write structural formulas of

a) the addition polymer formed from $H_2 C = \underset{}{\overset{\overset{H}{|}}{C}} - CN$

b) the condensation polymer formed between

$HO - CH_2 - CH_2 - OH$ and $HO - \underset{\underset{O}{\|}}{C} - CH_2 - \underset{\underset{O}{\|}}{C} - OH.$

- - - - - - - - - - - - - -

8. Consider the following organic compounds.

a) $H_3C - CH_2 - CH_2 - CH_3$

b) $H_3C - OH$

c) $H_3C - CH_2 - \overset{\overset{O}{\|}}{C} - OH$

d) $H_3C - \overset{\overset{H}{|}}{C} = CH_2$

e)

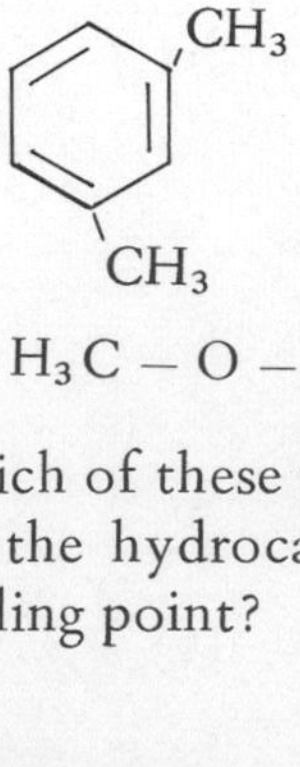

f) $H_3C - O - CH_3$

Which of these would you expect to show hydrogen bonding?
Of the hydrocarbons, which would you expect to have the highest boiling point?

Classify each of the oxygen compounds as to type.
For which of these compounds would you expect to find isomers with the same molecular formula?

9. Indicate whether each of the following substances is molecular, ionic, metallic, or macromolecular. Then, choose the member of each pair that you would expect to have the higher boiling point.

   a) KI or I_2 b) C or O_2 c) CO_2 or CS_2 d) Hg or He

10. Which of the following substances would you expect to be electrical conductors in the pure liquid state?

   a) Na   b) Cl_2 c) $MgCl_2$ d) SiO_2 e) NO_2 f) LiOH

11. Classify each of the following as paraffin, olefin, acetylene, or aromatic hydrocarbon.

   a) C_3H_8 b) C_4H_8 c) C_5H_8 d) C_7H_8

12. Draw structural formulas for each of the following compounds (include all isomers) and identify each structure as an alcohol, ether, aldehyde, ketone, acid, or ester.

   a) C_2H_4O b) $C_2H_4O_2$ c) C_2H_6O

13. Identify the monomers from which the following polymers were formed.

SELF-TEST

True or False

1. NaOH(s) is a poor electrical conductor because it is made up   (   )
of molecules.

2. The physical properties of molecular substances are directly ( )
related to the strengths of the covalent bonds holding the molecule
together.

3. Boiling points of molecular substances usually increase with ( )
molecular weight.

4. Ethyl alcohol, C_2H_5OH, would be expected to show ( )
hydrogen bonding.

5. The molecular formula C_5H_{10} represents an alkane. ( )

6. The saturated hydrocarbon C_4H_{10} shows *cis-trans* isomerism. ( )

7. The double bond in benzene and other aromatic hydro- ( )
carbons behaves chemically in the same way as that in ethylene.

8. An increase in pressure increases the stability of diamond ( )
relative to graphite.

Multiple Choice

9. Which one of the following solid substances consists of small, ( )
discrete molecules?
  a) graphite b) Dry Ice c) iron d) NaCl

10. Of the following interactions at the atomic-molecular level, ( )
the strongest is most likely to be
  a) dispersion   b) dipole
  c) hydrogen bond d) covalent bond

11. Which one of the following species could be boiled without ( )
breaking hydrogen bonds?
  a) CH_4 b) NH_3 c) HF d) none of these

12. What kind of force is *not* overcome in boiling a sample of ( )
water?
  a) hydrogen bond   b) dispersion force
  c) dipole force   d) polar covalent bond

13. Molecules themselves are broken apart during at least some ( )
phase changes in the case of
  a) SiC b) sulfur c) graphite d) all of these

14. In which pair of substances must the same type of attractive   ( )
force be overcome upon melting?
        a) LiCl and ICl   b) SiO_2 and CO_2
        c) CCl_4 and O_2 d) K and C

15. Which of the following would have a boiling point lower than   ( )
that of $SiCl_4$?
        a) $GeCl_4$ b) $SiBr_4$ c) CCl_4 d) LiCl

16. Which of the following is a gas at room temperature? (The   ( )
others are liquid or solid at room temperature.)
        a) C_4H_{10} b) C_5H_{12} c) C_3H_7OH d) $AlCl_3$

17. Compounds of carbon are so numerous and varied because   ( )
        a) carbon is the most abundant element in the universe
        b) carbon is the most abundant element on the earth's
           surface
        c) carbon atoms readily bond to one another
        d) there are more chemists studying organic compounds
           than inorganic compounds

18. Which of the following compounds is an alkene?   ( )
        a) $H_3C-CH_2-CH_3$ b) $H_2C=CH-CH_3$
        c) $HC\equiv C-CH_3$ d) $C_6H_5CH_3$

19. How many different compounds have the molecular formula   ( )
C_5H_{12} ?
        a) 1   b) 2   c) 3   d) some other number

20. Which of the following compounds is a ketone?   ( )
        a) $H_3C-\underset{H}{C}=O$ b) $H_3C-\underset{OH}{C}=O$
        c) $H_3C-\underset{CH_3}{C}=O$ d) $H_3C-\underset{OCH_3}{C}=O$

21. The compounds CH_3OH and $H_3C-COOH$ would react   ( )
together to produce
        a) an alcohol   b) a ketone   c) an ester   d) a fat

22. Which one of the following molecules might be expected to   ( )
form polymers with other molecules just like it?
        a) F_3C-CF_3 b) $F_2C=CF_2$ c) CF_4 d) F_2

23. The ability to form polymers depends on the presence of　( )
    a) carbon atoms
    b) unpaired electrons
    c) double bonds
    d) two at least potentially reactive sites on a molecule

24. Which substance, as a solid, should most readily conduct an　( )
electric current?
    a) C (graphite)　b) Ag　c) ice　d) quartz

25. Two-dimensional, or sheet-like, structures are observed in　( )
each of the following *except*
    a) nylon　b) graphite　c) diamond　d) mica

26. Ammonia, NH_3, is more soluble in ethyl alcohol　( )
($CH_3 CH_2 OH$) than is methane, CH_4. This is most likely a result of
    a) molecular weight difference
    b) difference in polarity
    c) a difference in densities
    d) hydrogen bonding

27. The various types of glass you use in the laboratory are　( )
unlikely to contain appreciable amounts of
    a) boron　b) hydrogen　c) oxygen　d) sodium

28. Color in a substance is often associated with　( )
    a) unpaired electrons
    b) alternating single and double bonds
    c) compounds of transitions metals
    d) all of these

29. Maple syrup has the property of being "syrupy" (viscous) as　( )
a result of
    a) being mostly water
    b) being more dense than water
    c) containing lots of protein polymers
    d) containing lots of alcohol functional groups, R—OH

30. Which one of the following should have the highest boiling ( )
point?

    a) $n-C_8H_{18}$
    b) $iso-C_8H_{18}$
    c) cyclooctane, C_8H_{16}

$$\begin{array}{c} H_3C \quad CH_3 \\ | \quad\quad | \\ d) \quad H_3C-C-C-CH_3 \\ | \quad\quad | \\ H_3C \quad CH_3 \end{array}$$

31. Large numbers of consumer products are derived from ( )
substances found in

    a) graphite   b) wood   c) glass   d) petroleum

SELF-TEST ANSWERS

1. **F** (Ions are fixed in position.)
2. **F** (Intermolecular forces are determining factors.)
3. **T** (With some exceptions, notably water.)
4. **T** (Compare H–O–H and –C–O–H.)
5. **F**
6. **F** (Double bond required.)
7. **F** (Contrast substitution vs addition.)
8. **T**
9. **b**
10. **d**
11. **a**
12. **d** (Molecules generally remain intact during phase changes; weaker intermolecular forces are overcome first.)
13. **d** (SiC and C are marcomolecular; cyclic S_8 molecules happen to break open easily.)
14. **c** (Dispersion forces.)
15. **c** (Lower molecular weight.)
16. **a** (Nonpolar, low molecular weight.)
17. **c** (d may be true, but irrelevant.)
18. **b**
19. **c** $$(C-C-C-C-C; \quad C-\overset{\overset{\textstyle C}{|}}{C}-C-C; \quad C-\overset{\overset{\textstyle C}{|}}{\underset{\underset{\textstyle C}{|}}{C}}-C)$$
20. **c**
21. **c**
22. **b** (Teflon is the commercial product.)
23. **d** (Either **b** or **c** may be involved, or two reactive functional groups.)

24. b
25. c (Three-dimensional.)
26. d (Both NH_3 and CH_3CH_2OH can hydrogen bond.)
27. b
28. d (Examples for a are O_2, NO_2; b, most dyes; c — see Chapter 19.)
29. d (Providing for lots of hydrogen bonding.)
30. a (Longest, least compact molecule.)
31. d (Particularly in the Age of Plastic.)

SELECTED READINGS

Amoore, J. E., and others, The Stereochemical Theory of Odor, *Scientific American* (February 1964), pp. 42-49.
   A summary of early work on a "lock and key" model based on the geometry of molecules, with several successful tests of the theory described.
Barrow, G. M., *The Structure of Molecules: An Introduction to Molecular Spectroscopy*, New York, W. A. Benjamin, 1963.
   A detailed examination of molecular energies and how they are experimentally determined and what they tell us about molecular structure.
Clapp, L. B., *The Chemistry of the OH Group*, Englewood Cliffs, N. J., Prentice Hall, 1967.
   An introduction to the properties of organic compounds, with an emphasis on correlation with structure.
Dence, J.B., Covalent Carbon-Metal(loid) Compounds, *Chemistry* (January 1973), pp. 6-13.
   Extends the descriptive chemistry of carbon to Groups 3A–5A.
Hall, S.K., Symmetry, *Chemistry* (March 1973), pp. 14–16.
   An introduction to the elements and operations of symmetry.
Materials, San Francisco, W. H. Freeman, 1967.
   The September 1967 issue of Scientific American. *This is a valuable introduction to the new science of materials (polymers, glasses, ceramics, etc.). This will be useful reading also for Chapter 9.*
Wagner, J. J., Nuclear Magnetic Resonance Spectroscopy — An Outline, *Chemistry* (March 1970), pp. 13-15.
   A qualitative introduction to one of the newer tools for determining molecular structure.

Liquids and Solids; Changes in State

QUESTIONS TO GUIDE YOUR STUDY

1. What properties do you generally associate with solids; liquids; gases? Can you qualitatively account for the differences?

2. What kinds of substances normally exist (i.e., at 1 atm, room temperature) as solids; liquids; gases?

3. Under what conditions may a given substance exist as a solid; liquid; gas?

4. Are there simple laws, as there are for gases, relating volume to temperature and pressure for a liquid or a solid?

5. How would you describe the process of evaporation; freezing? (First, in terms of what you observe; then in terms of atomic-molecular behavior.) What energy effects accompany these processes?

6. What kind of experiment would you do to show the existence of a vapor above a liquid? Or to show how the pressure of this vapor changes with temperature?

7. How would you recognize boiling? What would you measure?

8. How do we know the arrangements of molecules in solids and liquids? How do you account for the regularities of shape and size of crystalline solids? (Can you name some crystalline solids?)

9. How do you account for the fact that some solids are converted directly to gases, without first melting? (Examples: snow in the depths of winter; "dry ice".)

10. How are the properties of solids and liquids important to the activities of an architect; a ham radio operator; a man-eating shark?

11.

12.

YOU WILL NEED TO KNOW

Concepts

1. How many valence electrons there are in any given atom (a prediction that can be made from position in the periodic table) – Chapter 6
2. The ideas of kinetic theory about the nature of a gas (this chapter extends kinetic theory to liquids and solids) – Chapter 5

Math

1. How to use the ideal gas law, $PV = nRT$ – Chapter 5
2. How to find a logarithm or an antilogarithm (with either a log table or a slide rule) – See the text (Appendix 4), a math text, or the math preparation manual suggested in the Preface
3. How to plot the equation of a straight line; how to read such a graph

(e.g., $\log P = \dfrac{-\Delta H}{2.3\,RT} + B$ has the form $y = ax + b$) – See above (2)

4. The geometry of the cube: how to relate the dimensions – edge, body diagonal, face diagonal (For a cube with an edge of length ℓ: face diagonal $= \sqrt{2}\ell$, body diagonal $= \sqrt{3}\ell$.)

CHAPTER SUMMARY

The principal distinction between the gaseous state of matter, discussed in Chapter 5, and the condensed states (liquid and solid) considered in this chapter is the distance of separation between molecules. In the gas state, at ordinary temperatures and pressures, the molecules themselves account for a negligible fraction of the total volume. In the condensed states, the molecules ordinarily occupy from 50 to 70 per cent of the total volume. This structural difference explains why liquids and solids have much greater densities than gases, why they are less compressible, and expand less on heating.

Since molecules in the condensed states are closer together, short-range intermolecular forces exert a much greater influence on physical behavior. We have seen (Chapter 8) how the strength of these forces depends upon the type of molecule present. This is why we cannot write a simple equation of state, analogous to the ideal gas law, to describe the physical behavior of all liquids or all solids. Each liquid and solid has a characteristic density, compressibility, and expansibility.

The major distinction between the liquid and solid states is one of molecular mobility. The particles in a liquid are relatively free to move with respect to one another, while the particles in a solid are restricted to small vibrations about points in the crystal lattice. This makes the particle structure of solids much easier to study experimentally than that of liquids. The technique of x-ray diffraction can be applied to find the basic structural unit of the crystal, the unit cell; x-ray diffraction patterns for liquids are diffuse and difficult to interpret.

The particles in a solid tend to pack as closely as possible. (Why?) The metallic elements tend to crystallize in one of the two closest-packed arrangements (cubic or hexagonal). Another common type of packing is body-centered cubic, where the fraction of empty space is only a little greater. In ionic crystals, where the two kinds of ions differ in size, there is a greater variety of packing patterns. Frequently, the larger anions form a close-packed array, slightly expanded if necessary to accommodate the smaller cations fitting into "holes" in the anionic framework.

Much of this chapter is spent in discussing transitions from one state of matter to another. The nature of the equilibria involved can perhaps best be summarized by a phase diagram such as that shown on p. 251 of the text. At any given temperature, a liquid or solid has a characteristic vapor pressure (curves AB and AC, Figure 9.12). For all liquids and solids, vapor pressure rises exponentially with temperature; the general equation is:

$$\log_{10} \ P = \frac{-\Delta H}{(2.30) \ (1.99) \ T} + constant$$

where P is the vapor pressure, T the absolute temperature, and ΔH the enthalpy change for the transition (heat of vaporization or sublimation), in calories.

Two features of liquid-vapor equilibria which are not apparent from Figure 9.12 are the phenomena of boiling and critical behavior. A liquid boils (i.e., vapor bubbles form within the body of the liquid) when its vapor pressure becomes equal to the applied pressure; the normal boiling point is the temperature at which the vapor pressure becomes one atmosphere. A liquid can be heated in a closed container to temperatures far above its boiling point. Eventually, however, the kinetic energy of the molecules becomes too great for them to remain in the liquid; at this so-called "critical" temperature, the liquid suddenly and spontaneously vaporizes.

Again referring to the phase diagram on p. 251, we see that there is one particular temperature and pressure (the triple point) at which all three states of a substance are in equilibrium with each other. For most substances, the triple point pressure is relatively low. For a few substances, including CO_2, the triple point pressure exceeds one atmosphere. Such substances, when heated in an open container, pass directly from the solid to the vapor state (sublimate).

The line labeled AD in Figure 9.12 describes the equilibrium between the solid and liquid states. To the extent that this line deviates from the vertical, the melting point of the substance changes with pressure. For most substances, where the solid phase is the more dense, an increase in pressure favors the formation of solid and the melting point rises. For water, where the liquid is the more dense phase, the reverse is true. The effect is always small; pressures in the range of 50 to 100 atmospheres are required to change the melting point by as much as 1°C.

We need to keep in mind the ideas discussed back in Chapter 5 about the dynamic nature of matter at the atomic-molecular level. For example, the distribution of energies is central to our explanation of vapor pressure and its dependence on temperature. At any given temperature, a fraction of the molecules will have considerably greater than average energy — enough to escape from a liquid surface — while others are left waiting for collisions to impart the extra energy required to overcome intermolecular forces. Again, as the temperature is increased, a greater fraction of the molecules possess energy sufficient to break away from the rest of the condensed state, so that vapor pressure increases exponentially. Perhaps you can now see, at least qualitatively, how for a given substance (i.e., for a given magnitude of intermolecular forces, as measured by ΔH_{vap}) the temperature determines the vapor pressure.

BASIC SKILLS

1. Relate the heat of vaporization (heat of fusion) of a substance to the amount of heat required to boil (melt) a given mass of the substance.

- -

It is found that 0.0936 kcal of heat must be absorbed to vaporize one gram of liquid benzene, C_6H_6. What is the molar heat of vaporization of benzene? _____

Since one mole of benzene weighs 78.0 g, it is clear that the heat required to vaporize one mole of benzene must be 78.0 times that required to vaporize one gram. Setting up the calculation in a formal way:

$$1.00 \text{ mole } C_6H_6 \times \frac{78.0 \text{ g } C_6H_6}{1 \text{ mole } C_6H_6} \times \frac{0.0936 \text{ kcal}}{1.00 \text{ g } C_6H_6} = 7.30 \text{ kcal}$$

- -

The reverse calculation is required in Problem 9.5. Problem 9.24 illustrates the same principle; here, $-\Delta H = \Delta H_{vap} + \Delta H_{fus}$. In Problems 9.6 and 9.26, you must take into account the specific heat of water (1.00 cal/g°C) as well as its heat of fusion or vaporization.

2. **Have sufficient understanding of the concept of vapor pressure to work problems such as 9.7, 9.8, 9.26 and 9.27.**

This is an awkward way to phrase a "skill," but the only alternative is to resort to a lot of jargon. In certain parts of these problems, you have to apply the gas laws to calculate pressures or volumes (see also Example 9.1). You should keep in mind, however, that these laws can be applied only when there is a single, gaseous phase present. They cannot, for example, be used to calculate the pressure of a vapor in equilibrium with its liquid. That pressure is a characteristic physical property of a particular liquid at a given temperature, known as its equilibrium vapor pressure.

Pursuing this idea a bit further, a vapor at a pressure above its equilibrium vapor pressure is unstable; it will spontaneously condense until the pressure drops to the equilibrium vapor pressure. We cannot, for example, maintain steam at 100°C at a pressure above 1 atm. On the other hand, steam at 100°C at a pressure below 1 atm will remain at that pressure indefinitely, provided there is no liquid water around to establish equilibrium. (More about equilibrium in Chapters 12, 13 . . .)

3. **Use the Clausius-Clapeyron equation (Equation 9.4) to calculate one of the quantities P_2, T_2 or ΔH_{vap} given the values of the other two and P_1 at T_1.**

This skill is illustrated in Example 9.2. Here, the desired quantity is P_2, given T_2 (60°C), ΔH_{vap} (7300 cal/mole), P_1 (272 mm Hg), and T_1 (50°C). Problems 9.10 and 9.29a are analogous. In Problems 9.29b, the required quantity is T_2; you must also realize precisely what is meant by "normal boiling point."

Notice that in applying Equation 9.4, you must

- have at least a nodding aquaintance with logarithms (and anti-logarithms yet!);
- express temperatures in °K;
- use or obtain ΔH in cal/mole;
- express both pressures in the same units.

4. **Given the type of unit cell (simple cubic, face-centered cubic, or body-centered cubic), relate the cell dimensions to such quantities as atomic radius and density.**

The type of calculation involved here is illustrated, for a face-centered cubic unit cell, in Example 9.3. Problem 9.35 is entirely analogous to that example; Problem 9.16 applies the same reasoning to a body-centered cubic unit cell. The various relations involved in three types of unit cells are summarized in the table below.

	Simple	Face-Centered	Body-Centered
no. of atoms per cell	1	4	2
atoms touch along:	edge	face diagonal	body diagonal
relation between atomic radius (r) and cell length (ℓ)	$2r = \ell$	$4r = \ell\sqrt{2}$	$4r = \ell\sqrt{3}$

All the relationships just discussed apply to crystals in which only one kind of particle (e.g., a metal atom) is present. The same kind of geometric reasoning can be applied to ionic crystals, where we are dealing with two different kinds of ions. This application is covered in Problems 9.17 and 9.36.

5. Draw a phase diagram for a pure substance, given appropriate data, and state what phases are present at any given point on the diagram.

To illustrate what is involved here, consider Figure 9.12, p. 251. To draw this portion of the phase diagram for water, we might start by locating the triple point A at about 0°C and 5 mm Hg. Intersecting at point A are

- the vapor pressure curve, AB, for liquid water;
- the vapor pressure curve, AC, for ice;
- the melting point curve, AD, which tells us how the melting point of ice changes with the applied pressure.

To interpret the diagram, we note that within any given area such as that labelled "vapor" in Figure 9.12, only one phase is present. Along any line in the diagram, such as AB, two phases are in equilibrium. At the triple point, all three phases are present.

— —

Referring to Figure 9.12, what phases are present at B?_____at 25°C and 5 mm Hg?_____at C?_____at –5°C and 15 mm Hg? _____

Since B lies on the vapor pressure curve of the liquid, the two phases, liquid and vapor, must be in equilibrium. Similarly, the two phases solid and vapor must be in equilibrium at C. Locating 25°C and 5 mm Hg on the diagram, you should find that it is entirely within the area labelled "vapor"; only vapor is present. The point at –5°C and 15 mm Hg clearly lies within the solid region; only ice is present.

— —

Problems 9.18 and 9.37 require that you construct phase diagrams from information given in the statement of the problem (9.18) or scattered through various sections of the chapter (9.37).

6. Given the vapor pressure of a liquid at two or more temperatures, use Equation 9.3 to determine the heat of vaporization by a graphical method.

This problem perhaps arises most frequently as a laboratory exercise. To illustrate the principle involved, consider how we might determine the heat of vaporization of water from Figure 9.5, p. 239. To obtain the slope of the straight line, we may choose any two points. A convenient choice might be to pick the intersections of the line with the horizontal lines labelled 3.0 and 1.0 respectively. The x values corresponding to these intersections appear to be about "2.65" and "3.50" respectively. In other words:

$$y_2 = 3.00; \quad 10^3 x_2 = 2.65: \quad x_2 = 2.65 \times 10^{-3}$$

$$y_1 = 1.00; \quad 10^3 x_1 = 3.50: \quad x_1 = 3.50 \times 10^{-3}$$

$$\text{Slope} = \frac{y_2 - y_1}{x_2 - x_1} = \frac{3.00 - 1.00}{(2.65 - 3.50)10^{-3}} = \frac{2.00}{-0.85} \times 10^3 = -2.4 \times 10^3$$

$$\text{but, slope} = \frac{-\Delta H_{vap}}{(2.30)1.99}; \quad \Delta H_{vap} = -(2.30)(1.99)\text{slope}$$
$$= -(2.30)(1.99)(-2.4 \times 10^3)$$
$$= 11,000 \text{ cal}$$

Actually, this example is simplified by the fact that the plot has already been made for us. In Problems 9.9 and 9.28, you have to make your own plot and then carry out the type of analysis we have just gone through.

7. Use the Bragg equation (Equation 9.6) to relate the angle of diffraction θ and the wavelength of X-rays used, λ, to the distance between planes of atoms, d.

Problems of this type, of which 9.14 and 9.33 are typical, require little more than simple substitution into an equation. However, you will need trig tables or a slide rule with trig functions. This is perhaps the only time you will need such functions in the general chemistry course.

8. Use Trouton's Rule (Equation 9.5) to relate the molar heat of vaporization of a liquid to its normal boiling point.

- -

Given that diethyl ether boils at 35°C at 1 atm pressure, use Trouton's Rule to estimate its molar heat of vaporization. _____

From Equation 9.5, we see that $\Delta H_{vap} \approx 21 \times T_b$. Noting that T_b must be in $^\circ K$, we have:

$$T_b = (35 + 273)^\circ K = 308^\circ K$$

$$\Delta H_{vap} = (21 \times 308)cal = 6500 \text{ cal/mole}$$

The measured value (Table 9.2) is 6200 cal/mole.

- -

Problem 9.11 is entirely analogous to the example just worked; in Problem 9.30, one extra step is involved.

Problems

1. The heat of fusion of benzene, $C_6 H_6$, is 2.55 kcal/mole. How much heat must be absorbed to melt 10.0 g of benzene?
2. A sample of water vapor at 300 mm Hg and 100°C is cooled to 50°C at constant volume.

   a) Apply the Ideal Gas Law to calculate the pressure that will be exerted by the vapor at 50°C if no condensation occurs.
   b) Referring to the table of equilibrium vapor pressures on p. 659 of the text, state whether or not condensation will occur in (a). What will be the final pressure?

3. The vapor pressure of benzene is 75 mm Hg at 20°C; its heat of vaporization is 7300 cal/mole. Calculate the vapor pressure of benzene at 40°C.

4. Chromium crystallizes in a structure with a body-centered cubic unit cell 2.89 Å on an edge. Calculate the atomic radius of chromium.

5. Consider the phase diagram sketched below. State which phases are present at each of the numbered points. For this substance, is the density of the solid greater or less than that of the liquid?

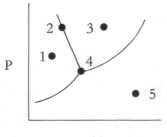

*6. Given the following data for water

t (°C)	20	40	60
vp (mm Hg)	18	55	149

a) For each point, calculate $\log_{10}$ (vp) and $1/T$, where T is in °K.
b) On a sheet of graph paper, plot $\log_{10}$ (vp) vs $1/T$.
c) Draw the best straight line through the three points.
d) Determine the slope of the line in (c).

e) Use the equation: slope $= \dfrac{-\Delta H_{vap}}{(2.30)\ (1.99)}$ to obtain the heat of vaporization.

*7. A beam of X-rays obtained by bombarding platinum with electrons has a wavelength of 0.182 Å. Calculate the angle of diffraction for n = 1, 2, and 3 when this X-ray beam strikes a crystal in which the layers of atoms are 1.60 Å apart.

*8. The heat of vaporization of liquid water is about 9700 cal/mole. Estimate what its normal boiling point would be in °C if Trouton's Rule were obeyed.

— — — — — — — — — — — — — —

9. A sample of ice at 0°C weighing 2.00 grams is converted to steam at 100°C.

a) How much heat is absorbed to melt the ice (ΔH_{fus} = 1.44 kcal/mole)?
b) How much heat is absorbed to raise the temperature of the water from 0°C to 100°C (specific heat of water = 1.00 cal/g°C)?
c) How much heat is absorbed to convert the water to steam at 100°C (ΔH_{vap} = 9.70 kcal/mole)?

10. A sample of benzene vapor at 80°C and 200 mm Hg is cooled at constant volume.

a) What will be the pressure at 50°C (vp benzene at 50°C = 272 mm Hg)?
b) What will be the pressure at 20°C (vp benzene at 20°C = 75 mm Hg)?

11. The heat of vaporization of a certain liquid is 6.22 kcal/mole. It has a vapor pressure of 52 mm Hg at 20°C.

   a) Calculate its vapor pressure at 50°C.
   b) Calculate the temperature at which the vapor pressure is 104 mm Hg.
   c) Calculate the normal boiling point, i.e., T at which vp = 1 atm.

12. Referring back to Problem 4:

   a) What is the volume of the unit cell in $(\text{Å})^3$? in cm^3 ?
   b) How many atoms should be assigned to each unit cell?
   c) What is the mass of the number of Cr atoms referred to in (b)? (AW Cr = 52.0)
   d) From your answers to (a) and (c), calculate the density of Cr in g/cm^3.

13. Construct a phase diagram for a substance given the following data:

   a) triple point: 20°C, 0.50 atm
   b) melting point: 22°C at 1.00 atm
   c) boiling point: 90°C at 1.00 atm
   d) vp of liquid: 0.70 atm at 70°C
   e) vp of solid: 0.30 atm at 0°C

   Show each of these points on the phase diagram.

- - - - - - - - - - - -

14. In an experiment described by W. L. Masterton and T. R. Williams, *J Chem Ed, 36,* 528 (1959), the molecular weight of a volatile liquid is determined by the use of the ideal gas law: M = gRT/PV. A sample of liquid is placed in a weighed flask, which is then placed in boiling water. The liquid is allowed to just completely evaporate, driving out the air and filling the flask (of volume V) with vapor at the temperature (T) of the boiling water and barometric pressure (P). The flask is stoppered and cooled so that the vapor condenses, then briefly opened so as to again establish barometric pressure within the flask. The flask is reweighed. In this weighing, however, some air has been replaced by vapor. (Why?) Now the molecular weight will be less than the true value, if it is calculated with g taken as simply the difference in the two weighings.
   Derive an expression which relates this erroneous reported value to the true value. What is the effect of the vapor pressure of the volatile liquid on the error?

SELF-TEST

True or False

1. At 50°C, the vapor pressure of liquid A is found to be 50 mm ( )
Hg; that of liquid B, 125 mm Hg. One can be reasonably sure that
liquid A has the higher normal boiling point.

2. For these same two liquids, one can state that the liquid with ( )
the higher surface tension at 50°C is probably A.

3. For any pure substance, the melting point is always a little ( )
above the freezing point. For instance, water will melt at a temperature above 0°C and freeze at a temperature a little below 0°C.

4. Not all solids are crystalline. ( )

5. Whenever bubbles form within a sample of liquid, the liquid ( )
is said to be boiling.

6. Generally, when a solid melts to form a liquid, the density ( )
decreases.

7. For all solids and liquids, vapor pressure increases linearly ( )
with temperature.

8. Below the triple point temperature, only solid can exist. ( )

9. Pure water cannot exist as a liquid below 0°C. ( )

Multiple Choice

10. When one mole of liquid water is placed in a 100 ml flask at ( )
25°C, it eventually establishes a constant pressure of 24 mm Hg. If a
200 ml flask were used instead, the final pressure would have been:
     a) 12 mm Hg b) 48 mm Hg c) 24 mm Hg d) 760 mm Hg

11. Which substance should have the highest vapor pressure at ( )
any given temperature?
     a) C_2H_6 b) C_3H_8 c) C_4H_{10} d) C_5H_{12}

12. The vapor pressure of CCl_4 at 77°C is 760 mm Hg. The heat ( )
of vaporization is approximately
     a) 1617 cal   b) 3790 cal   c) 7350 cal   d) 9340 cal

13. The boiling point of any liquid is: (   )
    a) 100°C
    b) the temperature at which as many molecules leave the liquid as return to it
    c) the temperature at which the vapor pressure is equal to the external pressure
    d) the temperature at which no molecules can return to the bulk of the liquid

14. The triple point is   (   )
    a) the temperature above which the liquid phase cannot exist
    b) usually found at a temperature very close to the normal boiling point
    c) the temperature at which the vapor pressure of a liquid is three times the value at 25°C
    d) the value of temperature and pressure at which solid, liquid, and gas may exist in equilibrium

15. A certain substance, X, has a triple point temperature of   (   )
20°C at a pressure of 2.0 atmospheres. Which one of the following
statements *cannot* be true?
    a) X can exist as a liquid above 20°C
    b) X can exist as a solid above 20°C
    c) liquid X is stable at 25°C and at 1 atm
    d) both liquid X and solid X have the same vapor pressure at 20°C

16. A certain solid sublimes at 25°C and 1 atmosphere pressure.   (   )
This means
    a) the solid is more dense than the liquid
    b) the pressure at the triple point is greater than 1 atmosphere
    c) the solid is less dense than the liquid
    d) the pressure at the triple point is less than 1 atmosphere

17. The melting point of benzene at 1 atmosphere is 5.5°C. The   (   )
density of liquid benzene is 0.90 g/ml; that of the solid is 1.0 g/ml.
At an applied pressure of 10 atm, the melting point of benzene
    a) is equal to 5.5°C
    b) is slightly greater than 5.5°C
    c) is slightly less than 5.5°C
    d) cannot be estimated from the information given

18. Dry ice, solid CO_2, is frequently used as a refrigerant; it ( )
undergoes sublimation at or near atmospheric pressure. This direct
conversion of solid CO_2 to gaseous CO_2 involves an absorption of
energy by the CO_2. This energy is used up, at least in part, in over-
coming
    a) covalent bonds   b) polar covalent bonds
    c) gravitational attractions   d) dispersion forces

19. When a solid melts to a liquid ( )
    a) greater attractive forces appear
    b) the molecules become more randomly oriented
    c) the molecules become less randomly oriented
    d) the container warms up

20. The number of closest neighbors in a body-centered cubic ( )
lattice is
    a) 4   b) 6   c) 8   d) 12

21. A certain metal crystallizes in a face-centered cubic unit cell. ( )
The relationship between the atomic radius (r) of the metal and the
length (l) of one edge of the cube is:
    a) $2r = l; r = l2$ b) $2r = (2)^{\frac{1}{2}}l; r = (2)^{\frac{1}{2}}l2$
    c) $4r = (2)^{\frac{1}{2}}l; r = (2)^{\frac{1}{2}}l/4$ d) $4r = (3)^{\frac{1}{2}}l; r = (3)^{\frac{1}{2}}l4$

22. A chemistry handbook normally gives the pressure at which ( )
the boiling point of a pure substance is measured but generally does
not indicate the pressure for the melting point. Why?
    a)   both melting point and boiling point are always measured
          at the same pressure
    b)   the melting point is usually nearly independent of
          pressure
    c)   solids are more often impure and therefore have variable
          melting points
    d)   all melting points are measured at one atmosphere

23. A sample of gaseous water at a pressure of 17.5 mm Hg is in ( )
equilibrium with liquid water at $20°C$. On reducing the volume
available to the system, you observe the pressure remains at 17.5 mm
Hg. The observation is accounted for by assuming:
    a)   gaseous water doesn't behave ideally under these
          conditions
    b)   some vapor escapes from the container
    c)   your pressure measurement was incorrect; it should have
          been higher
    d)   some vapor condenses to form liquid

24. For a liquid in equilibrium with its vapor, a straight line is ( )
obtained by plotting:
    a) P vs T b) log P vs T c) log P vs 1/T d) 1/P vs log T

25. Evacuating a container of liquid can cause   ( )
    a) boiling b) freezing c) evaporation d) all of these

26. A certain metal fluoride crystallizes in such a way that ( )
fluorine atoms occupy simple cubic lattice sites, while metal atoms
occupy the body centers of half the cubes. The formula of the metal
fluoride is:
    a) $M_2 F$ b) MF c) MF_2 d) MF_8

SELF-TEST ANSWERS

  1. **T** (At 50°C, B is closer to the boiling point than A since its vapor pressure is higher.)

  2. **T** (The lower v.p. would result from stronger intermolecular attractions; so would a higher surface tension.)

  3. **F** (They are one and the same equilibrium temperature — at which both solid and liquid can exist.)

  4. **T** (Many solids, such as glass and rubber, are *amorphous,* lacking the regularity of atomic placement found in crystals.)

  5. **F** (Bubbles often indicate dissolved gases coming out of solution.)

  6. **T** (Water is an exception.)

  7. **F** (Exponentially.)

  8. **F** (What about the pressure?)

  9. **F** (Supercooling! Also consider higher pressures.)

10. **c** (Vapor pressure depends only on temperature. Could all the water have evaporated?)

11. **a**

12. **c** ($\Delta H/T = 21$, approx.)

13. **c** (If you chose *b*, consider boiling water in an open container.)

14. **d**

15. **c** (Construct a phase diagram.)

16. **b**

17. **b**

18. **d** (Chapter 8.)

19. **b**

20. **c**

21. **c**

22. **b** (The solid-liquid line in a phase diagram of a pure substance is almost vertical.)

23. d
24. c
25. d (Evacuation will not only lower pressure, at least temporarily, but also will remove the "hottest" molecules.)
26. c (One F atom per cube; ½ M atom per cube.)

SELECTED READINGS

Apfel, R.E., The Tensile Strength of Liquids, *Scientific American* (December 1972), pp. 58-71.
Of negative pressures and how liquids tear — more properties of liquids.

Bernal, J. D., The Structure of Liquids, *Scientific American* (August 1960), pp. 124-128. *Much of the work on liquid structure has been done by the author. He makes comparisons to other states and presents a model for a state which we commonly, carelessly, think of as being without structure.*

Bragg, L., X-Ray Crystallography, *Scientific American* (July 1968), pp. 58-70.
A detailed survey of the techniques and successes of x-ray diffraction, written by one who has made history in the field. About the same level as the text.

Fullman, R. L., The Growth of Crystals, *Scientific American* (March 1955), pp. 74-80.
An interesting, well-written introduction to the nature of crystal growth: the "screw-dislocation" theory.

Greene, C. H., Glass, *Scientific American* (January 1961), pp. 92-105.
Undercooled liquid or inorganic polymer? The article describes the structure of this noncrystalline solid and the processes by which it is made.

Hittinger, W.C., Metal-Oxide-Semiconductor Technology, *Scientific American* (August 1973), pp. 48-57.
About the properties and processing of solid state devices essential to modern computers and pocket calculators.

Moore, W. J., *Seven Solid States: An Introduction to the Chemistry and Physics of Solids*, New York, W. A. Benjamin, 1967.
Often advanced, this is a discussion of seven exemplary solids, including salt, gold, ruby and steel. Much material assembled here.

Sanderson, R. T., The Nature of 'Ionic' Solids, *J. Chem. Ed.* (September 1967), pp. 516-523.
A different approach to the solids we have labeled as bonded ionically — that emphasizes the connection with covalent bonding. At least read the summary.

Turnbull, D., The Undercooling of Liquids, *Scientific American* (January 1965), pp. 38-46.
Supercooling is a general phenomenon. This article discusses "nucleation" and the nature of the freezing process, as well as reviews the structure of liquids.

Wells, A. F., *The Third Dimension in Chemistry*, New York, Oxford, 1956.
This book will let you find some use for that geometry you learned long ago. Also of interest might be the huge volume Structural Inorganic Chemistry, same author, same publisher, authoritative.

10

Solutions

QUESTIONS TO GUIDE YOUR STUDY

1. What kinds of processes occur within solids; within liquids? For example, what happens at the molecular level when two liquids are mixed; or when a solid dissolves in a liquid?

2. How do the properties of a substance (such as f.p., b.p.) change when a second substance is dissolved in it? Are there simple laws relating solution properties?

3. Can you think of some naturally occurring solutions? (Recall that we generally do not encounter pure substances outside the lab.)

4. How do you quantitatively describe the composition of a solution?

5. How do the properties of a solution depend on the relative amounts of its components?

6. What properties of substances determine the extent to which one will dissolve in the other? What generalizations can be made?

7. How do changes in conditions, such as temperature and pressure, affect solubility?

8. How would you experimentally show that a given solution is saturated; supersaturated? How would you prepare such solutions?

9. Can you describe the structure, the bonding, of solutions? Can you account for any energy effects accompanying the formation of a solution?

10. To what uses are solutions put in the laboratory?

11.

12.

YOU WILL NEED TO KNOW

Concepts

1. How to predict the geometry and polarity of molecules; how to predict bond type and relative magnitudes of intermolecular forces — Chapter 7

Math

1. How to express amounts of substances in the various common units: grams, moles, number of particles — Chapter 3

CHAPTER SUMMARY

We saw in Chapter 9 that the magnitude of a substance's intermolecular forces determines whether it exists as a gas, liquid, or solid at normal temperatures and pressures (e.g., 25°C, 1 atm). Again, we might expect two substances with intermolecular forces of about the same magnitude to be infinitely soluble in each other. Putting these two generalizations together, we expect complete miscibility only if the two substances involved are in the same physical state. Experiment confirms this reasoning: all gases are infinitely soluble in one another; complete miscibility is the rule rather than the exception among liquids of similar polarity. In contrast, solids and gases always show limited solubility in liquid solvents. Carrying this reasoning one step further, we deduce that the closer a solid or gas is to the liquid state (i.e., the lower the melting point of the solid or the higher the normal boiling point of the gas) the greater will be its solubility. At the opposite extreme, solutes whose intermolecular forces differ vastly from those of the solvent will be virtually insoluble in it (e.g., the permanent gases and hydrocarbons in water).

It is, of course, possible to change the solubility of a solute by changing the external conditions of temperature and pressure. If the solution process is endothermic, an increase in temperature promotes solubility. This is almost always the case with solid-liquid systems and usually true when both components are liquids. With a gas, the solution process may be either endothermic or exothermic; the solubility of the permanent gases in organic solvents normally increases with temperature while the same gases become less soluble in water when the temperature is raised. Solubility is little affected by pressure except for gas-liquid systems where it is directly proportional to the partial pressure of the gas over the solution.

Various methods are used to express the concentrations of solutions, i.e., the relative amounts of solute and solvent. Frequently, the weight fractions are given by quoting weight per cents of the two components or, for very dilute solutions, "parts per million" or even "parts per billion" of solute (1 ppm = 1 g solute per 10^6 g solvent). In the laboratory, reagent concentrations are most frequently expressed in terms of molarity (no. of moles per liter of solution). Two other concentration units, mole fraction and molality, are particularly useful for expressing the colligative properties of solutions. The terminology here is unfortunate; students often confuse molality (m) with molarity (M), or even with morality, which is something else altogether.

Almost without exception, the addition of a small amount of solute lowers the freezing point of a solvent; if the solute is nonvolatile, the boiling point will be raised as well. Both of these effects can be related to the lowering of solvent vapor pressure that always accompanies the formation of a solution. One can also explain the phenomenon of osmosis in terms of vapor pressure lowering: water or other solvent moves through a semipermeable membrane from a region of high vapor pressure (pure solvent) to one of low vapor pressure (solution). Osmosis can be prevented by exerting sufficient force on the solution side of the membrane; the pressure required to do this is referred to as the osmotic pressure.

Each of the effects just described (vapor pressure lowering, freezing point depression, boiling point elevation, osmotic pressure) is a colligative property; its magnitude depends primarily upon the concentration of solute particles rather than their type. The appropriate equations appear in the text (Equations 10.8, 10.9, 10.10, 10.11). Notice that these equations incorporate a constant of proportionality which in every case but one is characteristic of the particular solvent. The exception is the expression for osmotic pressure, where the gas law constant R appears. (This constant keeps popping up in the most unexpected places!) These equations can be used in a very practical way to calculate vapor pressures, freezing points, boiling points, and osmotic pressures of solutions. In addition they serve as the basis for the experimental determination of the molecular weight of a solute. (How would you rationalize the similarity of the two equations: $\pi V = nRT$ and the Ideal Gas Law? In what ways is a dilute solution like a gas?)

BASIC SKILLS

1. Relate the masses of the components of a solution to:

   a) the weight percentages of the components;
   b) the mole fractions of the components;
   c) the molality of the solute.

The application of these skills is illustrated in Example 10.1 (weight percentages), Example 10.2 (mole fractions) and Example 10.3 (molality).

Note that each example is worked directly from the defining equation for the concentration unit (Equations 10.1 to 10.3).

To practice these skills, try Problems 10.3a, b; 10.4b, c, d; 10.22a; 10.23a, b, c. Other parts of these problems involve the concept of molarity (see below) or, in one case, the use of the Ideal Gas Law.

2. Given or having calculated two of the three quantities: molarity, number of moles of solute, number of liters of solution, obtain the other quantity.

This skill is shown in Example 10.4. In part (a), you are given the volume of solution and its molarity and must calculate first the number of moles of solute and then the number of grams. In part (b), the required quantity is the volume of solution; the number of moles and molarity are given.

Simple calculations involving molarity as a concentration unit are required in Problems 10.4a, 10.5, and 10.24a. These problems are analogous to Example 10.4 in that you need only substitute into the defining equation (Equation 10.4) to obtain the desired quantity. Problem 10.23d is slightly more involved in that you must first obtain the volume of solution (note that the density is given).

3. Relate the volumes and molarities of solutions prepared by dilution with solvent.

No new skill is involved here. All you need do is recall the definition of molarity and realize that diluting a solution with solvent does not change the number of moles of solute. Refer to Example 10.5 and Problem 10.6. In Problem 10.25, you are asked to derive a general equation which could be used to solve any problem of this type.

4. Use Henry's Law (Equation 10.6) to relate the solubility of a gas in a liquid to its partial pressure.

The solubility of O_2 in water at 20°C and one atmosphere is 1.38×10^{-3} mole/liter. What is its solubility at the same temperature but at a partial pressure of 0.20 atm? _____

Since solubility is directly proportional to partial pressure, we can write

$$C_2 = C_1 \times \frac{P_2}{P_1}$$

where C_2 and C_1 refer to solubilities at P_2 and P_1 respectively. Substituting:

$$C_2 = 1.38 \times 10^{-3} \frac{\text{mole}}{\text{liter}} \times \frac{0.20 \text{ atm}}{1.00 \text{ atm}} = 2.8 \times 10^{-4} \frac{\text{mole}}{\text{liter}}$$

Note that in any calculation of this sort, C_2 and C_1 must be expressed in the same units; the same is true for P_2 and P_1.

_ _

Example 10.6 illustrates a slightly different approach to this type of problem, where the constant k is solved for explicitly. Problem 10.8 offers a simple example of a Henry's Law calculation. Problem 10.27 is entirely analogous to Example 10.6, with the addition of a final step in which one goes from concentration to amount of N_2.

5. Predict the relative solubilities of two different solutes in a given solvent, or of a given solute in two different solvents.

There is a simple principle involved here: *solubility is increased by similarity in the type and strength of intermolecular forces in solute and solvent.* The implications of this principle, some of which are not entirely obvious, are illustrated in the following example.

_ _

At 25°C and 1 atm, which would be the more soluble in a nonpolar liquid solvent such as benzene: $H_2O(l)$ or $C_6H_{14}(l)$? _____ H_2 (g, bp = -253°C) or CH_4 (g, bp = -183°C)? _____ naphthalene (s, mp = 80°C) or anthracene (s, mp = 218°C)? _____

In the first case, the obvious choice is hexane, C_6H_{14}, since its intermolecular forces are of the same type (dispersion) as those in benzene. Water, which is hydrogen-bonded, is only very slightly soluble in nonpolar solvents. With the other pairs, the type of intermolecular force (dispersion) is the same in both cases. The choice for the more soluble species is the solute whose intermolecular forces are nearest in *magnitude* to those of the liquid solvent. These are the gas, CH_4 and the solid, naphthalene, which are closest to the liquid state at room temperature. Gases which are difficult to condense (H_2) and solids which are difficult to melt (anthracene) have intermolecular forces quite different in magnitude from those of a liquid such as benzene and hence have relatively low solubilities.

_ _

Problems 10.9, 10.10, and 10.28 illustrate this principle; in certain parts of these problems you are expected to recognize the physical state of the solute. Problem 10.29 is essentially similar but a little more thought-provoking.

6. Use Raoult's Law to relate:

a) the vapor pressure of solvent in a solution to its mole fraction;
b) the vapor pressure lowering to the mole fraction of solute;
c) the total vapor pressure of a solution to the mole fractions of solute and solvent.

The appropriate equations for these calculations are given on pp. 276-7 of the text. The calculations for (b) and (c) are illustrated in Example 10.7 and the paragraph that follows. To illustrate (a), consider the following example.

_ _

What is the vapor pressure of water in a solution containing 100 g of sugar, $C_{12}H_{22}O_{11}$, in 500 g of water, H_2O, at 25°C? _____

In order to apply Equation 10.7, p. 276, we need to know the vapor pressure of pure water at 25°C (23.76 mm Hg) and the mole fraction of water in the solution. Since the molecular weights of H_2O and $C_{12}H_{22}O_{11}$ are 18.0 and 342 respectively:

$$X_1 = \frac{\text{no. moles } H_2O}{\text{no. moles } H_2O + \text{no. moles } C_{12}H_{22}O_{11}} = \frac{500/18.0}{500/18.0 + 100/342} = 0.990$$

$$P_1 = X_1 P_1{}^0 = (0.990)(23.76 \text{ mm Hg}) = 23.5 \text{ mm Hg}$$

_ _

Problems 10.11, 10.12, 10.30 and 10.31 all apply Raoult's Law to vapor pressure calculations.

7. Use Equations 10.9, p. 278 (boiling point elevation) and 10.10, p. 279 (freezing point lowering) to obtain:

   a) the boiling point or freezing point of a nonelectrolyte solution.
   b) the molecular weight of a nonelectrolyte.

The calculations involved in (a) are illustrated in Example 10.8; those required to obtain the molecular weight of a solute are shown in Example 10.9. Note that:

— in order to work problems of this type, you must be thoroughly familiar with the concept of molality, the concentration unit used in Equations 10.9 and 10.10.
— these equations apply only to nonelectrolytes (i.e., molecular solutes); electrolyte solutions are considered in Chapter 11.

Problems 10.13 and 10.32 further illustrate skill 7(a); Problems 10.14 and 10.33 are similar in principle, but require additional conversions. Problems 10.17 and 10.18 involve the calculation of molecular weight from freezing point lowering. In Problems 10.36 and 10.37 an additional step is involved; the molecular formula is obtained by combining the molecular weight and the simplest formula (Chapter 3).

8. Use Equation 10.11, p. 283, to relate the osmotic pressure of a solution, π, to the molarity of the solute, M.

- -

What is the osmotic pressure of an aqueous solution containing 100 g of sugar, $C_{12}H_{22}O_{11}$, per liter of solution at 25°C? _____

To use Equation 10.11, we must first calculate the molarity of the sugar. Since the molecular weight of $C_{12}H_{22}O_{11}$ is 342, we have:

$$M = \frac{\text{no. moles } C_{12}H_{22}O_{11}}{\text{no. liters solution}} = \frac{100/342 \text{ mole}}{1.00 \text{ liter}} = 0.292 \text{ mole/liter}$$

Also: $R = 0.0821$ lit atm/mole°K; $T = (273 + 25)°K = 298°K$

Hence: $\pi = (0.292)\ (0.0821)\ (298)$ atm $= 7.14$ atm

- -

Problems 10.15 and 10.34 offer a straightforward application of Equation 10.11. Problem 10.16 is a little more difficult and considerably more interesting. Osmotic pressure measurements, like freezing point and boiling point determinations, can be used to obtain molecular weights of nonelectrolytes (Problems 10.19 and 10.38).

Problems

1. A student prepares a solution by dissolving 12.0 g of methyl alcohol, CH_3OH, in 20.0 g of water, H_2O.

   a) What are the weight percentages of CH_3OH and H_2O in this solution?
   b) What are the mole fractions of CH_3OH and H_2O?
   c) What is the molality of CH_3OH?

2. What is the molarity of a solution prepared by dissolving 109 g of $BaCl_2$ to give 2.60 liters of solution?

3. What is the molarity of a solution prepared by diluting 200 ml of 0.100 M NaCl to 500 ml?

4. The solubility of argon in benzene at 25°C and 0.15 atm pressure is 0.0075 mole/liter. What is its solubility at 25°C and one atm?

5. Which is the more soluble in water:

a) $CH_3OH(l)$ or $C_5H_{12}(l)$?
b) $He(g, bp = -269°C)$ or $Ar (g, bp = -186°C)$?
c) solid A (mp = 310°C) or B (mp = 120°C), assuming A and B have similar structures?

6. In a certain solution, the mole fractions of benzene and toluene are 0.20 and 0.80 respectively.

a) What is the vapor pressure of benzene above the solution $(P^0 = 76$ mm Hg$)$?
b) What is the vapor pressure of toluene above the solution $(P^0 = 24$ mm Hg$)$?
c) What is the total vapor pressure above the solution?

7. a) Calculate the freezing point and boiling point (at 760 mm Hg pressure) of a solution containing 12.0 g of urea, $CO(NH_2)_2$, in 150 g of water.
b) A solution of 0.500 g of a certain nonelectrolyte in 10.0 g of water freezes at $-0.365°C$. What is the molecular weight of the nonelectrolyte?

8. What is the osmotic pressure at 30°C of a solution containing 0.240 mole of solute in 6.00 liters of solution?

– – – – – – – – – – – –

9. A solution is prepared by adding enough water to 50.0 g of glucose, $C_6H_{12}O_6$, to form one liter of solution. The density of the solution is 1.020 g/ml.

a) What is the molarity of glucose?
b) What is the total mass of the solution?
c) What is the mass of water in the solution?
d) What is the weight percentage of glucose in the solution? of water?
e) What is the molality of glucose?
f) What is the mole fraction of glucose? of water?

10. Using your answers to (9), calculate the freezing point, boiling point at 760 mm Hg, osmotic pressure at 20°C, and vapor pressure lowering at 25°C of the solution.

11. Complete the following table for water solutions.

Solute	no. of grams	no. of moles	molarity	no. of liters soln.
NaCl	13.0	____	1.65	____
NaCl	____	2.43	____	6.50
$CaCl_2$	____	____	0.250	2.00
$CaCl_2$	15.0	____	____	4.00

12. It is desired to prepare 5.00 liters of a 0.400 M solution by diluting a 1.00 M solution.

a) What volume of 1.00 M solution should be used?
b) What volume of water should be added? (Assume no change in volume on mixing.)

13. For each of the following pairs of substances, predict which will be the more soluble, first in benzene and then in water.

a) N_2 (g, bp = –196°C) or O_2 (g, bp = –183°C)
b) NaCl or CCl_4
c) Naphthalene (mp = 80°C) or 2-methyl naphthalene (mp = 35°C)
d) H_2O_2 or C_6H_{14}

14. One gram samples of solutes A, B and C are dissolved in 10.0 grams of water to give three separate solutions. Use the information below to calculate the molecular weights of A, B and C.

a) Solution A freezes at –0.518°C.
b) Solution B boils at 100.27°C at 760 mm Hg.
c) Solution C has an osmotic pressure of 1.60 atm at 25°C (in the equation for π, take molarity to be equal to molality).

SELF-TEST

True or False

1. In a solution containing equal numbers of grams of benzene   (   )
(MW = 78) and toluene (MW = 92), the mole fractions are each 0.50.

2. A supersaturated solution of air in water could be prepared   (   )
by bubbling air through water at 60°C and cooling to 25°C.

3. The mole fractions of all the components of a solution add to   (   )
unity.

4. If water saturated with nitrogen at 1 atm is exposed to air, ( )
N_2 will come out of solution.

5. DDT should be more soluble in alcohol than in carbon tetra- ( )
chloride.

6. The solubility of a solid in a liquid is ordinarily directly ( )
proportional to the absolute temperature.

7. Raoult's Law often applies to the solvent in a solution but ( )
not to the solute.

8. The boiling point of a 1 m water solution of a nonelectrolyte ( )
will be 100.52°C, provided the solution is ideal.

9. Freezing point lowering, boiling point elevation, and osmotic ( )
pressure can all be explained in terms of vapor pressure lowering.

10. The freezing point of a 1 m solution of K_2SO_4 would be ( )
about the same as that of a 1 m solution of urea.

Multiple Choice

11. When one mole of KCl is dissolved in a kilogram of water, the ( )
concentration of Cl^- is:
     a) 0.5 molal   b) 0.5 molar   c) 1.0 molal   d) 1.0 molar

12. A student wishes to prepare 100 cc of 0.50 M NaCl from ( )
2.00 M NaCl. What volume of the more concentrated solution should
he start with?
     a) 25 cc   b) 50 cc   c) 100 cc   d) 400 cc

13. The solubility of a certain salt in water is 22 g/liter at 25°C ( )
and 60 g/liter at 80°C. A student prepares 500 cc of saturated
solution at 80°C, cools to 25°C, and adds a tiny crystal of salt. How
many grams of salt come out of solution?
     a) 11   b) 19   c) 22   d) 8

14. The number of grams of Na_2SO_4 (FW = 142) required to ( )
prepare 2.00 liters of 1.50 M solution is:
     a) 3   b) 213   c) 142   d) 426

15. To obtain 12.0 g of K_2CrO_4 from a solution labeled "5.0% ( )
K_2CrO_4 by weight," how many grams of solution should you weigh
out?
     a) 2.4   b) 5.0   c) 12   d) 240

16. Which of the following has the least effect on the solubility ( )
of a solid in a liquid solvent?
    a) nature of solute   b) nature of solvent
    c) temperature   d) pressure

17. The solubility of CO_2 in water at 20°C is 3.4 g/liter at 1 atm. ( )
If the CO_2 pressure is raised to 3 atm, the solubility, in g/liter, is
expected to be
    a) 3.4/3   b) 3   c) 3.4   d) 3.4 × 3

18. For which of the following pairs would you expect solubility ( )
to be the greatest?
    a) O_2-water b) sugar-water c) sugar-benzene d) O_2-N_2

19. A student wants to remove naphthalene, an aromatic hydro- ( )
carbon, from his lab coat. What solvent would you recommend?
    a) water   b) ethyl alcohol   c) benzene   d) sulfuric acid

20. When one mole of a nonvolatile nonelectrolyte is dissolved in ( )
two moles of a solvent, the vapor pressure of the solution, relative to
that of the pure solvent is:
    a) 1/3   b) 1/2   c) 2/3   d) cannot tell

21. Which one of the following is not a colligative property? ( )
    a) freezing point   b) molarity
    c) osmotic pressure   d) color

22. The osmotic pressure of a 1 M solution of sugar at room ( )
temperature would be closest to:
    a) 0.25 atm   b) 1 atm   c) 2.5 atm   d) 25 atm

23. To determine the molecular weight of a polymer with an ( )
approximate molecular weight of 10,000, one would probably
measure the
    a) osmotic pressure of a solution
    b) boiling point of a solution
    c) density of the solid
    d) vapor pressure of a solution

24. A small amount of a certain solute which melts at 800°C is ( )
added to water. The solution will be expected to freeze
    a) above room temperature   b) slightly above 0°C
    c) at 0°C   d) slightly below 0°C

25. In a 0.1 M solution of NaCl in water, which one of the ( )
following will be closest to 0.1?
    a) mole fraction NaCl   b) mole fraction water
    c) wt. % NaCl   d) molality

26. A solution of 1.00 g of a nonelectrolyte in 20.0 g of water ( )
freezes at –0.50°C. The molecular weight of the nonelectrolyte is:

a) 1.86/(0.50) (0.020)   b) 1.86/(0.50) (20.0)
c) 0.50 (20.0)/1.86   d) 0.020(0.50)/1.86

SELF-TEST ANSWERS

1. **F** (Larger no. moles of benzene.)
2. **F** (Solubility increases as temperature drops; it wouldn't even be saturated at the lower temperature.)
3. **T**
4. **T** (P_{N_2} above the solution is reduced.)
5. **F**
6. **F** (Relationship not linear.)
7. **T**
8. **F** (Depends on pressure, solute volatility.)
9. **T**
10. **F** (More solute particles with K_2SO_4.)
11. **c**
12. **a**
13. **b** (I.e., 30–11.)
14. **d**
15. **d** (I.e., 12.0 g K_2CrO_4 × 100 g soln./5.0 g K_2CrO_4.)
16. **d**
17. **d**
18. **d** (Same physical state.)
19. **c** (d would remove the lab coat.)
20. **c**
21. **d**
22. **d**
23. **a** (Most sensitive.)
24. **d** (Nonvolatile solute lowers vapor pressure and mp.)
25. **d**
26. **a**

SELECTED READINGS

Dye, J. L., The Solvated Electron, *Scientific American* (February 1967), pp. 76-83.
For many years a lab curiosity, it now appears that the solvated electron may be an intermediate reactive species in many reactions.
Mysels, K. J., The Mechanism of Vapor-pressure Lowering, *J. Chem. Ed.* (April 1955), p. 179.
Points out errors in the argument usually given (but doesn't offer anything in its place).

For working problems pertaining to solutions (concentrations, reaction stoichiometry, colligative properties, etc.) see problem manuals listed in the Preface.

Water, Pure
and Otherwise

QUESTIONS TO GUIDE YOUR STUDY

1. Where and in what physical state do you find water, near and not so near the earth?

2. Is pure water, like many other pure substances, found only in the laboratory?

3. What's so special about water? For example: How does the heat required for changes in state compare to that of other substances? How does the melting point compare to that for a substance of similar molecular weight and geometry? How do its solvent properties compare to those of other liquids?

4. What geometry and bonding do you predict for the water molecule? How does this structure help explain water's unique properties?

5. How do water solutions compare to nonaqueous solutions? Do the same laws of colligative properties and the same principles of solubility apply?

6. What solutes are found in naturally occurring water solutions? What are their sources?

7. How can you measure the concentration of a solute in water solution?

8. How can you prepare very pure water? How is drinking water normally prepared? (What processes are used in a municipal water plant?)

9. Do we have, as we do for mixtures of gases, a workable model for simple salt solutions?

10. Can you justify an extensive study of water solutions? (The remaining chapters in the text are devoted almost entirely to reactions which occur in water solution.)

11.

12.

YOU WILL NEED TO KNOW

Concepts

1. The kinds and relative strengths of forces that exist between molecules — Chapter 8
2. How to write and interpret chemical equations — Chapter 3

Math

1. How to work problems involving concentration units — Chapter 10
2. How to perform stoichiometric calculations (those based on chemical equations) — Chapter 3

CHAPTER SUMMARY

The unique properties of water (e.g., high boiling point and heat of vaporization, volumetric behavior near $0°C$) are related to the ability of the H_2O molecule to form three-dimensional networks held together by hydrogen bonds. Such networks are known to exist in ice, accounting for its low density relative to liquid water. They are believed to persist, probably in modified form, in liquid water near the melting point. As the temperature is raised, there is a shift toward a more closely packed arrangement typical of normal liquids. Qualitatively at least, we can explain the unusual behavior of water near $0°C$ (minimum volume at $4°C$) in terms of an equilibrium between "flickering clusters" of hydrogen bonded water molecules and a more closely packed structure.

Despite a great deal of research in the area, the effect of electrolytes on the water structure is still a matter of controversy. At low concentrations, below 0.01 m, the freezing point lowering of a solution of a 1:1 electrolyte such as NaCl can be expressed by the equation:

$$\Delta T_f = i\,(1.86°C)\,m; \quad i = 2 - 0.78\,m^{1/2}$$

If the solute ions (e.g., Na^+, Cl^-) behaved as completely independent particles, i would be equal to 2. We see from this equation that as we go to very, very dilute solutions ($m \to 0$), where the ions are extremely far apart, $i \to 2$. The correction term in the expression for i takes account of interactions between oppositely charged ions, the "ion-atmosphere" effect of Debye. Equations analogous to this can be written for other types of electrolytes; we always find that as $m \to 0$, i approaches the number of moles of ions per mole of electrolyte. All of these equations break down at embarrassingly low concentrations; nevertheless, they can be applied to freezing point data in very dilute solutions to obtain i. In this way, it is possible to deduce the way in which a species ionizes in water: we can demonstrate for example that H_3PO_4 forms H^+ and $H_2PO_4^-$ ions (i = 2), while H_2SO_4 gives $2 H^+ + SO_4^{2-}$ (i = 3).

Relatively small amounts of suspended or dissolved species in water can make it unsafe for drinking, unappealing for recreational purposes, or harmful in other ways to the environment. One measure of the extent of pollution of water by organic material is its B.O.D. (biochemical oxygen demand), which tells us the concentration of oxidizable organics in water. In their tendency to be oxidized, even perhaps to the harmless products CO_2 and H_2O, these pollutants will consume (make a demand on) dissolved oxygen. Vapor phase chromatography (Chapter 1) can be used to determine pesticides which concentrate in the food chain to the point where they have become a menace to wildlife and perhaps to human beings as well. Many inorganic contaminants can have undesirable effects. An example is the phosphate ion, present in many detergents, which perhaps contributes to the eutrophication of lakes. Compounds of certain of the heavy metals, notably mercury and lead, are known to be toxic even at very low concentrations.

Among the processes commonly used in municipal water purification are sedimentation, coagulation, filtration, and disinfection. These are designed primarily to clarify water and destroy disease-carrying organisms. Waters containing relatively high concentrations of Ca^{2+}, Mg^{2+} or Fe^{3+} can be softened by passing through columns containing natural or synthetic zeolites, which replace the objectionable cations by Na^+ ions. Alternatively, the same result can be achieved by the lime-soda process in which Ca^{2+} ions are precipitated as $CaCO_3$.

Salty or brackish waters can be purified by a variety of processes. The one in most common use today is the age-old process of distillation. Ion exchange columns containing two resins, one which exchanges H^+ ions for cations, the other OH^- ions for anions, can also be used. Still another approach which shows considerable promise is reverse osmosis, in which sufficient pressure is applied to the salt solution to cause water to move out of the solution through a membrane to a reservoir of pure water. The recycling of water is a very real problem: it has been estimated that the available water resources in the U.S. are daily replenished by some 4×10^{11} gallons; the current rate of water use has been estimated to be 4×10^{11} gallons!

BASIC SKILLS

This chapter is primarily descriptive; the new concepts introduced are largely qualitative in nature (e.g., water structure, water pollution, water purification). Most of this material lends itself to discussion and informed speculation rather than formal problem solving. You will note that most of the problems at the end of the chapter require discussions or simple conversions. A few quantitative ideas are presented; the more important are considered below.

1. Given the simplest formula of an electrolyte, calculate the limiting value of i in Equations 11.2 to 11.4, p. 296.

This amounts simply to finding the number of moles of ions produced by the complete dissociation of one mole of electrolyte. It is illustrated in the following examples; see also Problem 11.4.

Electrolyte	no. moles cation	no. moles anion	i
KCl	1	1	2
$MgSO_4$	1	1	2
$CaCl_2$	1	2	3
K_2SO_4	2	1	3
$Al(NO_3)_3$	1	3	4
Na_3PO_4	3	1	4
$Al_2(SO_4)_3$	2	3	5

2. Use Equations 11.2 to 11.4, p. 296, to estimate:

a) ΔT_f, ΔT_b or π for an electrolyte solution given or having calculated i and m;

b) the nature of ionization of an electrolyte, given ΔT_f, ΔT_b or π at a known m.

- -

Estimate the freezing points of 0.10 molal solutions of KCl and K_2SO_4, taking the limiting values for i.

Applying Equation 11.2 to KCl (i = 2) and K_2SO_4 (i = 3), we have:

KCl: $\Delta T_f = 2(1.86°C)(0.10) = 0.37°C$; $T_f = -0.37°C$

K_2SO_4: $\Delta T_f = 3(1.86°C)(0.10) = 0.56°C$; $T_f = -0.56°C$

Because of ion interactions, the freezing point lowerings observed are somewhat less than those just calculated (see Table 11.4).

_ _

Skill 2(b) is illustrated in Example 11.1. See also Problems 11.9 and 11.25 (the latter is entirely analogous to part b of Example 11.1).

3. Calculate the BOD of a water sample, given either

    a)   data for the amount of oxygen consumed by a water sample.
    b)   the amount of oxidizable material present in a water sample.

Skill 3(a) is illustrated in Example 11.2. See also Problem 11.10, which is entirely analogous; Problem 11.26 is similar. Skill 3(b) is illustrated in the following example.

_ _

The only oxidizable organic matter present in a one liter sample of water is 1.5 grams of acetic acid, CH_3COOH. Given the equation for its reaction with oxygen:

$$CH_3COOH(aq) + 2O_2(aq) \rightarrow 2CO_2(aq) + 2H_2O$$

what is the BOD of the sample? _____

In effect, we are asked to calculate the number of milligrams of O_2 required to react with 1.5 g of acetic acid. This is really a simple problem in stoichiometry, of the type discussed in Chapter 3. Noting that one mole of CH_3COOH (60.0 g) reacts with two moles of O_2 (2 × 32.0 g = 64.0 g), we have:

$$\text{no. mg } O_2 = 1.5 \text{ g } CH_3COOH \times \frac{64.0 \text{ g } O_2}{60.0 \text{ g } CH_3COOH} \times \frac{10^3 \text{ mg}}{1 \text{ g}} = 1.6 \times 10^3 \text{ mg } O_2$$

The BOD must then be 1.6×10^3 mg O_2/liter water.

_ _

Problem 11.11 is analogous to the example just worked, except that the volume of sample is not one liter. In Problem 11.27, you are required to carry out the reverse calculation. That is, given the BOD, find out how much CN^- must be present.

4. Given the concentrations of Ca^{2+} and HCO_3^- in a water sample of known volume, calculate how much $Ca(OH)_2$ and Na_2CO_3 should be added to soften it.

As indicated in the discussion on pp. 312-3, the most economical way to soften water by the lime-soda process is to

a) add one mole of $Ca(OH)_2$ per two moles of HCO_3^-; this removes one mole of Ca^{2+} from the hard water;

b) remove any Ca^{2+} remaining by adding Na_2CO_3 in a 1:1 mole ratio.

This principle is applied in Example 11.3 and in Problems 11.14 and 11.30. (In case you are curious as to why SO_4^{2-} and Cl^- ions appear in some of these problems, it may be well to point out that solutions have to be electrically neutral. For example, a solution containing 1.6×10^{-4} mole/liter of Ca^{2+} and no other ions would "soften" itself spontaneously via a spectacular electric discharge.)

Problems

1. Give the limiting value of i for

    a) KNO_3 b) $CoCl_2$ c) $Fe_2(SO_4)_3$ d) $(NH_4)_2CO_3$ e) H_2SO_4

2. Calculate the freezing points of 0.01 molal solutions of each of the electrolytes in (1), assuming complete dissociation.

3. Consider the reaction between urea and oxygen:

    $$2CO(NH_2)_2 (aq) + 3O_2 (g) \rightarrow 2N_2 (g) + 4H_2O + 2CO_2 (g)$$

    A certain water supply contains one kilogram of urea in 10^4 liters of water.

    a) How many grams of O_2 are required to react with the urea?
    b) What is the BOD of the water?

4. How many moles of $Ca(OH)_2$ and Na_2CO_3 should be added to soften one liter of water if the concentrations of Ca^{2+} and HCO_3^- are respectively:

    a) 1.2×10^{-4} M and 2.4×10^{-4} M
    b) 1.2×10^{-4} M and 1.6×10^{-4} M

- - - - - - - - - - - -

5. Estimate the osmotic pressure of a 0.10 M solution of $K_2Cr_2O_7$ at 30°C, assuming the limiting value of i.

6. A certain weak acid is partially disociated in water according to the equation

$$HA(aq) \rightleftharpoons H^+(aq) + A^-(aq)$$

The freezing point of a 0.20 molal solution is $-0.45°C$.

a) Calculate the experimental value of i.
b) What would be the value of i if the HA were 0% dissociated? 100% dissociated? 50% dissociated?
c) Calculate the % dissociation in this solution.

7. Consider the reaction between oxygen and urea referred to in Problem 3. If the BOD is found to be 52 mg O_2 /liter:

a) how many mg of urea are there per liter?
b) what is the concentration of urea in g/liter?
c) what is the molarity of urea?

8. How many moles of $Ca(OH)_2$ and Na_2CO_3 should be added to soften ten liters of water in which the concentrations of various ions are:

	Na^+	Ca^{2+}	HCO_3^-	Cl^-
a)	1.0×10^{-4} M	2.2×10^{-4} M	_____	5.4×10^{-4} M
b)	_____	1.1×10^{-4}	1.6×10^{-4}	0.6×10^{-4}
c)	1.2×10^{-4}	_____	_____	1.2×10^{-4}
d)	1.2×10^{-4}	_____	0.6×10^{-4}	0.6×10^{-4}

- - - - - - - - - - - - -

9. Suppose we wish to remove the five-day biochemical oxygen demand of 140 from a sample of municipal waste by adding oxygen via air-saturated water. For each liter of waste, how much water should we add? By what factor is the waste water diluted with the air-saturated water? At 25°C and one atmosphere, the solubility of oxygen in water is 1.2×10^{-3} molar.

SELF-TEST

True or False

1. The difference in strength between the hydrogen bonds in H_2O and those in HF explains why the solid phases of these compounds have very different structures. ( )

2. The maximum density of water at $4°C$ can be explained by assuming that the molecular structure of ice does not completely disappear when it melts. ( )

3. A 0.1 m solution of KCl should freeze at about $-0.19°C$. ( )

4. It is believed that the $(C_2H_5)_4N^+$ ion promotes the formation of "flickering clusters" of water molecules. Consequently, one would expect the maximum density of solutions containing this ion to be above $4°C$. ( )

5. A high B.O.D. suggests that a water supply is contaminated with organic wastes. ( )

6. Branched-chain detergents are more readily broken down in nature than those with straight chains. ( )

7. In the lime-soda process of water softening, Ca^{2+} ions in the hard water are replaced eventually by Na^+ ions. ( )

8. A zeolite column used to soften water could be regenerated by flushing with a concentrated solution of $CaCl_2$. ( )

9. Salt water can be converted to pure water by the process of osmosis. ( )

Multiple Choice

10. The orientation of covalent and hydrogen bonds about an oxygen atom in ice is best described as ( )
    a) bent   b) planar   c) tetrahedral   d) hexagonal

11. The property of water that is most critical to the regulation of body temperature is ( )
    a) high specific heat   b) high boiling point
    c) high heat of vaporization   d) high surface tension

12. At low concentrations, i (Equation 11.2) for $Fe(NO_3)_3$ approaches ( )
    a) 1   b) 2   c) 3   d) 4

13. For a 0.1 m solution of KNO_3, i would probably be ( )
   a) about 1   b) slightly less than 2
   c) slightly greater than 2   d) about 5

14. An anion which apparently promotes the growth of algae in ( )
water is
   a) Cl^- b) SO_4^{2-} c) PO_4^{3-} d) CN^-

15. The principal advantage of an insecticide like Sevin over DDT ( )
is that it
   a) is less expensive
   b) is less toxic to humans
   c) remains effective longer
   d) breaks down more quickly

16. Mercury-containing organic compounds resemble DDT in ( )
that they
   a) have similar toxicities
   b) usually arise from the same source
   c) accumulate in the food chain
   d) have similar molecular structures

17. Which one of the following processes does not remove ( )
suspended matter from water?
   a) sedimentation   b) coagulation
   c) filtration   d) chlorination

18. One way to remove the unpleasant taste of chlorinated water ( )
is to
   a) filter it   b) add NH_3 c) add NaOH   d) soften it

19. A certain water supply contains Ca^{2+} and HCO_3^- in a 1:2 ( )
mole ratio. To soften this water by the lime-soda process, one should
add
   a) only lime b) only soda c) both lime and soda d) CO_2

20. In reverse osmosis, the relationship between the applied ( )
pressure P and the osmotic pressure π would most likely be
   a) $P < \pi$ b) $P = \pi$ c) $P > \pi$ d) $P \gg \pi$

21. Which one of the following substances would be most likely ( )
to increase the BOD of a water supply?
   a) CO_2 b) O_3 c) $C_6H_{12}O_6$ d) N_2

22. Which of the following is generally *not* considered a water ( )
pollutant?
   a) CN^- b) Fe^{3+} c) Ca^{2+} d) all are pollutants

23. What happens to the amount of dissolved oxygen in a stream   (   )
as the temperature rises?
     a) it increases   b) it decreases
     c) it remains unchanged   d) cannot say

SELF-TEST ANSWERS

   1. F (Tetrahedral coordination is key factor.)
   2. T
   3. F (Closer to $-0.37°C$; limiting value of $i = 2$.)
   4. T (Clusters break down more slowly as T rises.)
   5. T
   6. T
   7. T
   8. F (Use NaCl.)
   9. F (Reverse osmosis.)
10. c
11. c (Would you expect a to be important too?)
12. d (Four ions/mole: $Fe^{3+} + 3 NO_3^-$.)
13. b (Attractive forces between ions effectively reduce their number.)
14. c
15. d
16. c
17. d
18. b
19. a (Add one mole of $Ca(OH)_2$ for every two moles of HCO_3^-.)
20. c (If $P \gg \pi$, the membrane might break.)
21. c
22. d (Depends on intended use; otherwise, b or c might be acceptable.)
23. b (One of the more drastic effects of thermal pollution.)

SELECTED READINGS

Arrhenius, S., Über die Dissociation der in Wasser gelösten Stoffe, *Zeitschrift für physikalische Chemie*, Vol. I, pp. 631-648 (1887).
    Exercise your scientific German by reading Arrhenius' theory of electrolytic solutions. Begins with a brief summary of earlier work done by Clausius and van't Hoff.
Cleaning Our Environment: The Chemical Basis for Action, Washington, D. C., American Chemical Society, 1969.
    A survey of problems and programs and some of the chemistry involved in pollution. A useful handbook to start with.
Derjaguin, B. V., Superdense Water, *Scientific American* (November 1970), pp. 52-64.
    A discussion of the controversial "anomalous" water of Problem 11.1, since denied.

Frank, H. S., The Structure of Ordinary Water, *Science* (August 14, 1970), pp. 635-641.
A generally readable elaboration of the "flickering clusters" model discussed in this chapter.

Goldwater, L. J., Mercury in the Environment, *Scientific American* (May 1971), pp. 15-21.
A calm, balanced analysis of the general topic — mercury in the environment: sources and effects.

Montague, K. and P., *Mercury,* San Francisco, Sierra Club, 1971.
A rather biased discussion written in a popular, journalistic style. Worth contrasting with the above article.

Runnels, L. K., Ice, *Scientific American* (December 1966), pp. 118-126.
A surprising number of great scientists have worked on the problem of the structure of ice and its properties (as well as those of liquid water). This is an excellent article on this work: thorough and reasonably up-to-date.

Stoker, H. S. and S. L. Seager, *Environmental Chemistry: Air and Water Pollution,* Glenview, Ill., Scott, Foresman, 1972.
Lots more facts and figures here than almost anywhere else. A lot of chemistry applied in a straightforward manner. No obvious bias.

Turk, A. and others, *Ecology, Pollution, Environment,* Philadelphia, W. B. Saunders, 1972.
A very readable introduction to ecological problems of pollution. There isn't very much chemistry applied here; the problems are often not as simple to analyze as the authors suggest.

12

Spontaneity of Reaction; ΔG and ΔS

QUESTIONS TO GUIDE YOUR STUDY

1. What do you mean by reaction *spontaneity*? (What is commonly meant by the word spontaneous?) What special sense is implied here?

2. Can you decide whether or not a particular reaction will occur without even carrying out the reaction? (Recall being able to calculate ΔH for such a case.)

3. What physical meaning do you associate with ΔG and ΔS (and ΔH)? What kinds of measurements can you make to determine their values?

4. How are ΔG and ΔS (and ΔH) related to the masses of substances taking part in a reaction?

5. How are these quantities related to each other (and to ΔH)? What can be said about the interrelation (interconversion?) of various forms of energy? How is each of these forms related to the properties of atoms and molecules, like bond energy and molecular geometry?

6. How do ΔG and ΔS (and ΔH) depend on reaction conditions such as temperature and pressure? Can you predict the sign or magnitude of the effect of a change in conditions for ΔG; ΔS; and ΔH?

7. Can you predict the sign or relative size of ΔS for a given reaction? Likewise for ΔG?

8. What happens at the molecular level when the entropy of a system increases; when free energy decreases? (What is the sign convention here? Recall that a positive ΔH means that heat is transferred *to* the system.)

9. To what kinds of systems or reaction conditions are the principles of spontaneity discussed in this chapter applicable; not applicable?

10. Are you now able to decide the conditions under which any given reaction may occur? (Of what significance or usefulness is all this to the nonchemist or to society?)

163

11.

12.

YOU WILL NEED TO KNOW

Concepts

1. How to interpret the sign of ΔH; and, in general, how to write and interpret "thermochemical equations" – Chapter 4

Math

1. How to calculate ΔH for any reaction – Chapter 4

CHAPTER SUMMARY

If you were to ask several nonscientists to define a "spontaneous" process, you might get a variety of answers. One response might be, "a process taking place by itself without anyone having to work to bring it about." This statement resembles the definition of spontaneity presented in this chapter, where we say that (at constant temperature and pressure) *a spontaneous process is one which is capable of producing work.* Notice that it is the inherent capacity to do work that characterizes a spontaneous reaction. We say that the combustion of methane at 25°C and 1 atm is spontaneous because this reaction, if carried out in an appropriate machine, will produce energy in the form of work. The fact that under normal conditions the energy released when methane burns is dissipated as heat does not alter our conclusion.

The capacity of a reaction to produce work can be related to a fundamental property of substances known as free energy. The difference in free energy, ΔG, between products and reactants is a direct measure of the maximum amount of work that can be obtained from a reaction. Spontaneous reactions are ones for which the products have a lower free energy than the reactants (ΔG < 0). If the free energy of the products is greater than that of the reactants (ΔG > 0), work must be done to make the reaction go and we say that it is nonspontaneous. If, perchance, the free energies of reactants and products are equal (ΔG = 0), the reaction system is balanced on a knife-edge; a tiny "push" will cause it to go in one direction or the other.

We can think of the free energy change as being made up of two components; according to the Gibbs-Helmholtz equation:

$$\Delta G = \Delta H - T\Delta S$$

The enthalpy change, ΔH, which we became acquainted with in Chapter 4, represents the amount of heat absorbed or evolved when a reaction is carried out at constant pressure. If the bonds in the product molecules are stronger than those in the reactants, ΔH will be negative and the reaction will be exothermic. The Gibbs-Helmholtz equation tells us that a negative value of ΔH will tend to make ΔG negative as well. We conclude that, other things being equal, exothermic reactions will tend to be spontaneous.

The other quantity appearing in this equation, ΔS, represents the difference in entropy between products and reactants. Entropy is a measure of randomness. The entropy of a solution exceeds that of the pure components; gases have greater entropies than liquids or solids. We note from the Gibbs-Helmholtz equation that a positive value of ΔS will tend to make ΔG negative and hence contribute to spontaneity. This analysis confirms what experience tells us: order degenerates into chaos without any help from anyone.

The balance between these two factors depends upon the magnitude of the absolute temperature, T. At low temperatures, $T\Delta S$ will be small and the sign of ΔG will be that of ΔH. As temperature rises, the entropy factor plays a more important role and eventually dominates. If, as most often happens, ΔH and ΔS have the same sign, the direction in which a reaction proceeds spontaneously will reverse at some temperature. A simple example is the vaporization of water at 1 atm pressure (ΔH and ΔS both positive). Below 100°C, ΔH is greater than $T\Delta S$, ΔG is positive, and vaporization does not occur. At 100°C, $\Delta H = T\Delta S$, $\Delta G = 0$, and the system is at equilibrium. Above 100°C, ΔS predominates, $\Delta G < 0$, and water at atmospheric pressure boils spontaneously.

Of the three quantities in the Gibbs-Helmholtz equation: 1) ΔH is essentially independent of both temperature and pressure; 2) ΔS can be taken to be temperature independent, at least above room temperature, but is strongly dependent upon pressure for many reactions; and 3) ΔG is ordinarily dependent upon both temperature and pressure. Throughout this chapter, we have confined our discussion to reactions taking place at 1 atm and hence have dealt almost exclusively with $\Delta G^{1\,atm}$. This quantity is often calculated from free energy of formation of compounds.

$$\Delta G^{1\,atm} = \Sigma \Delta G_f^{1\,atm} \text{ products} - \Sigma \Delta G_f^{1\,atm} \text{ reactants}$$

Electrochemical measurements are generally the experimental source for these quantities (Chapter 21).

BASIC SKILLS

1. Given a table of standard free energies of formation, calculate $\Delta G^{1 \text{ atm}}$ at 25°C for a reaction.

The calculation here is entirely analogous to that used in Chapter 4 to obtain ΔH for a reaction from heats of formation. Since ΔG, unlike ΔH, varies considerably with both temperature and pressure, these two conditions must be specified, i.e., we calculate "$\Delta G^{1 \text{ atm}}$ at 25°C."

The skill is illustrated in Example 12.1 and applied directly in Problem 12.6. Problem 12.20 shows how the free energy of formation of a compound can be calculated from ΔG for one of its reactions. In the example and in both of these problems, the significance of the sign of ΔG is emphasized. ΔG is negative for a spontaneous reaction and positive for a nonspontaneous reaction.

Several of the problems at the end of Chapter 12 require as a first step the calculation of ΔG from free energies of formation (Problems 12.11 to 12.14 and 12.25 to 12.28). In several cases, it is also necessary to calculate ΔH from the table of heats of formation in Chapter 4.

2. Estimate the sign of ΔS for certain physical and chemical changes.

The general principle here is that in going from a more ordered to a more random state, entropy increases. Examples of such processes are fusion, vaporization, and the formation of a solution. For chemical reactions, we can be sure that ΔS will be positive if there is an increase in the number of moles of gas. For reactions which do not involve gases, ΔS is usually positive if there is an increase in the number of moles.

————————————————————————————————————

Predict the sign of ΔS for each of the following processes.
$H_2O(l) \rightarrow H_2O(g)$ _____
$H_2O(l) + CH_3OH(l) \rightarrow$ solution _____
$2H_2(g) + O_2(g) \rightarrow 2H_2O(g)$ _____

For the first two processes, we would expect ΔS to be positive since we are going from a more ordered state (pure liquid) to a more random state (gas or solution). For the chemical reaction, ΔS should be negative, since there is a decrease in the number of moles of gas (3 moles gas → 2 moles gas).

————————————————————————————————————

See Problems 12.8 and 12.22, which include applications of this principle to areas outside the natural sciences (part b of each problem).

3. Given the heat of vaporization (heat of fusion) and normal boiling point (freezing point) of a substance, calculate $\Delta S^{\,1\,atm}$ for the process.

This calculation is illustrated in Example 12.3. Note that in order to obtain $\Delta S^{\,1\,atm}$, we must use the normal boiling point or freezing point, i.e., the temperature at which the two phases are at equilibrium at 1 atm pressure. Only at that temperature is $\Delta G^{\,1\,atm} = 0$ and hence $\Delta S^{\,1\,atm} = \Delta H/T$.
   Problems 12.9 and 12.23 apply this skill.

4. Given or having calculated ΔH and $\Delta G^{\,1\,atm}$ at 25°C, use the Gibbs-Helmholtz equation (Equation 12.14) to obtain:

   a) $\Delta S^{\,1\,atm}$: $\Delta S^{\,1\,atm} = (\Delta H - \Delta G^{\,1\,atm}$ at 25°C)/298° K
   b) $\Delta G^{\,1\,atm}$ at any temperature T: $\Delta G^{\,1\,atm} = \Delta H - T\Delta S^{\,1\,atm}$
   c) the temperature at which the reaction is at equilibrium at one atmosphere pressure: $T = \Delta H/\Delta S^{\,1\,atm}$

These several applications of the Gibbs-Helmholtz equation are illustrated in Examples 12.4 and 12.5. The example below illustrates the combined applications for a single reaction.

_ _

For a certain reaction, $\Delta H = +19.0$ kcal and $\Delta G^{\,1\,atm}$ at 25°C = +15.0 kcal. What is:

   a) $\Delta S^{\,1\,atm}$?
   b) $\Delta G^{\,1\,atm}$ at 1000°K?
   c) T at which the reaction is at equilibrium at 1 atm pressure?

Applying the above equations:

$\Delta S^{\,1\,atm} = (\Delta H - \Delta G^{\,1\,atm}$ at 25°C)/T = (19.0 kcal - 15.0 kcal)/298°K = 0.013 kcal/°K

$\Delta G^{\,1\,atm}$ at 1000°K = $\Delta H - 1000(\Delta S^{\,1\,atm}) = 19.0$ kcal -13.0 kcal = +6.0 kcal

$$T = \frac{19.0 \text{ kcal}}{0.013 \text{ kcal/}^\circ\text{K}} \approx 1500^\circ \text{K}$$

_ _

Note that the quantities required for this calculation (ΔH and $\Delta G^{\,1\,atm}$ at 25°C) are ordinarily obtained from tables of heats and free energies of formation. This is the case with all the problems of this type at the end of

Chapter 12 (12.11 to 12.14; 12.25 to 12.28). It may also be worth pointing out that we do not really need to know $\Delta G^{1\ atm}$ *at 25°C* to obtain $\Delta S^{1\ atm}$. If, for example, we were given $\Delta G^{1\ atm}$ at 500° K, we could obtain $\Delta S^{1\ atm}$ as:

$$\Delta S^{1\ atm} = (\Delta H - \Delta G^{1\ atm}\ at\ 500°\ K)/500°\ K$$

Problems

1. Using Table 12.2, p. 328, calculate $\Delta G^{1\ atm}$ at 25°C for the reactions:

   a) $CuO(s) + H_2(g) \rightarrow Cu(s) + H_2O(l)$
   b) $2KClO_3(s) \rightarrow 2KCl(s) + 3O_2(g)$

2. Predict the sign of ΔS for each of the following processes:

   a) $C_6H_6(l) \rightarrow C_6H_6(s)$
   b) $2NO(g) + O_2(g) \rightarrow 2NO_2(g)$
   c) $2AgCl(s) \rightarrow 2Ag(s) + Cl_2(g)$

3. Calculate $\Delta S^{1\ atm}$ for the fusion and vaporization of Hg (ΔH_{fus} = 560 cal/mole, ΔH_{vap} = 14,200 cal/mole, normal melting point = –39°C, normal boiling point = 367°C).

4. For a certain chemical reaction, ΔH = –21.0 kcal and $\Delta G^{1\ atm}$ at 25°C = –9.0 kcal. Calculate:

   a) $\Delta S^{1\ atm}$ b) $\Delta G^{1\ atm}$ at 700° K   c) T at which $\Delta G^{1\ atm}$ = 0

- - - - - - - - - - -

5. Consider the reaction: $N_2(g) + 3H_2(g) \rightarrow 2NH_3(g)$

   a) Calculate ΔH for this reaction (Table 4.1, p. 70).
   b) Calculate $\Delta G^{1\ atm}$ at 25°C (Table 12.2, p. 328).
   c) Calculate $\Delta S^{1\ atm}$ and explain why it has a negative sign.
   d) Calculate the temperature at which $\Delta G^{1\ atm}$ = +10.0 kcal.

6. Suppose the reaction in Problem (5) were: $N_2(g) + 3H_2(g) \rightarrow 2NH_3(l)$. Given that the heat of vaporization of NH_3 is +5.6 kcal/mole, its normal boiling point is –33°C, and using the answers to (5) as a starting point, calculate

   a) ΔH b) $\Delta S^{1\ atm}$ c) $\Delta G^{1\ atm}$ at 25°C

- - - - - - - - - - -

7. Of the following reactions, show by calculation which method would be best for preparing iron at the lowest temperature.

$$Fe_2 O_3 (s) \rightarrow 2 Fe(s) + 3/2 O_2 (g)$$

$$Fe_2 O_3 (s) + 3/2 C(s) \rightarrow 2 Fe(s) + 3/2 CO_2 (g)$$

$$Fe_2 O_3 (s) + 3 H_2 (g) \rightarrow 2 Fe(s) + 3 H_2 O(g)$$

8. In the combustion of carbon and many of its compounds at or near atmospheric pressure, one usually observes the formation of CO_2. Show by calculation why one expects CO_2 and not CO as the usual combustion product. (You might wish to choose a particular compound, or carbon itself, or perhaps try to relate CO and CO_2 more directly.) Try to give a molecular level interpretation to your answer. Would you expect CO to be the major product at higher temperatures? Again, justify your answer by means of a calculation.

9. Certain metals, when heated with excess Cl_2 (g), form two or more different chlorides, depending on the temperature. Invariably, it is the chloride containing the *lowest* percentage of chlorine which is formed at the highest temperatures. (e.g., MCl_2 forms rather than MCl_3.) Choose any metal that forms at least two chlorides and illustrate by appropriate calculation the validity of the above statement. Convenient tabulations of thermodynamic data may be found in the handbooks listed in the Preface to this guide.

SELF-TEST

True or False

1. At 25°C and 1 atmosphere pressure, the reaction between N_2 and O_2 to form NO(g) is spontaneous. ( )

2. For the decomposition of water to the elements at 25°C and 1 atm, $\Delta G = +56.7$ kcal. This means that at least 56.7 kcal of work has to be supplied to make this reaction go. ( )

3. If ΔG for a reaction is positive, it is impossible to carry out the reaction unless either the temperature or the pressure is changed. ( )

4. "Work," as the term is used in this chapter, includes radiant energy. ( )

5. Exothermic reactions always become spontaneous as we approach absolute zero. ( )

6. Free energies of formation of compounds are positive numbers in most cases. ( )

7. The free energies of formation of Cu_2O and CuO at 25°C and 1 atm are −35 and −30 kcal/mole respectively. This means that Cu_2O, exposed to oxygen at room temperature, will convert spontaneously to CuO. ( )

8. For the reaction: $PCl_5(g) \rightarrow PCl_3(g) + Cl_2(g)$, ΔS is positive. ( )

9. Automobiles increase in entropy with age. ( )

10. If ΔH and ΔS for a reaction are both negative, we expect ΔG to be negative at all temperatures. ( )

11. Reactions for which ΔH and ΔS have the same sign will tend to reverse at high temperatures. ( )

12. For the process $CO_2(s) \rightarrow CO_2(g)$, we expect both ΔH and ΔS to be positive. ( )

13. For the process referred to in (12), ΔS should decrease with increasing pressure. ( )

Multiple Choice

14. For the reaction at 25°C and 1 atm: ( )
$$CH_4(g) + 2O_2(g) \rightarrow CO_2(g) + 2H_2O(l),$$
which one of the following statements is *not* true?
  a) the reaction is exothermic
  b) the reaction is spontaneous
  c) ΔG < 0
  d) work has to be done to make the reaction go

15. Which one of the following quantities is independent of pressure? ( )
  a) ΔG   b) ΔH   c) ΔS   d) W_{max}

16. The free energy of formation of AgCl is −26.2 kcal/mole. ΔG for the reaction: $2\ AgCl(s) \rightarrow 2\ Ag(s) + Cl_2(g)$ is: ( )
  a) −52.4 kcal b) −26.2 kcal c) +26.2 kcal d) +52.4 kcal

17. Which one of the following statements best describes the ( )
relationship between ΔG and temperature?
 a) ΔG is independent of T
 b) ΔG varies with T
 c) ΔG is a linear function of T
 d) ΔG usually decreases with T

18. Which of the following would you expect to have the largest ( )
entropy per mole?
 a) Li(s) b) Li(g) c) $LiCl \cdot H_2O(s)$ d) $Cl_2(g)$

19. The heat of fusion of benzene is 2.55 kcal/mole. Its melting ( )
point is 5°C. ΔS for the melting of benzene, in cal/mole°K, is about
 a) 0.5   b) 2.6   c) 9   d) 13

20. For which of the following would you expect ΔS to be ( )
nearest zero?
 a) $C(s) + O_2(g) \rightarrow CO_2(g)$
 b) $2\,SO_2(g) + O_2(g) \rightarrow 2\,SO_3(g)$
 c) $CaSO_4(s) + 2\,H_2O(l) \rightarrow CaSO_4 \cdot 2\,H_2O(s)$
 d) $CO(g) + \frac{1}{2}\,O_2(g) \rightarrow CO_2(g)$

21. For a certain reaction, ΔH is 20 kcal and ΔG at 25°C is 14 ( )
kcal. ΔS in cal/mole°K is approximately:
 a) +6000   b) −6000   c) +20   d) −20

22. Vaporization is an example of a process for which: ( )
 a) ΔH, ΔS and ΔG are positive at all temperatures
 b) ΔH and ΔS are positive
 c) ΔG is negative at low T, positive at high T
 d) ΔH is strongly pressure dependent

23. For a certain reaction, $\Delta H = +2.5$ kcal, $\Delta S^{1\ atm} = 10$ cal/ ( )
mole°K. This reaction will be at equilibrium at 1 atm at about
 a) 0.25°C   b) 25°C   c) −23°C   d) cannot tell

24. The free energy of formation of CO is –32.8 kcal/mole at ( )
25°C; its heat of formation is –26.4 kcal/mole. As the temperature is
increased, ΔG_f will
 a) remain unchanged
 b) change sign
 c) become less negative
 d) become more negative

25. When aniline dissolves in hexane, ΔS is positive. Why then is ( )
aniline only slightly soluble in hexane at room temperature? Choose
the best thermodynamic explanation.
   a) an equilibrium is reached, at which point, $\Delta G = 0$
   b) ΔH is positive
   c) the intermolecular forces in the two liquids are quite
      different
   d) the formation of a solution is always a spontaneous
      process

26. When the city of New York burns coal in an electrical ( )
generating plant, the entropy of the universe
   a) increases   b) decreases   c) is not changed   d) is zero

27. An emission control device in a 1973 model automobile ( )
absorbs gasoline vapors onto solid activated carbon. During the
spontaneous process of adsorption,
   a) the carbon becomes warm
   b) the carbon becomes cool
   c) there is no change in T
   d) the T may change, but one cannot say how

SELF-TEST ANSWERS

1. F (This is suggested by the co-existence of the two gases in the
      atmosphere, and proved by the fact that ΔG_f of NO is positive.)
2. T
3. F (See question 2; nonspontaneous does not mean impossible.)
4. T (Radiant energy brings about the nonspontaneous process of
      photosynthesis.)
5. T (ΔH determines the sign of ΔG at low T.)
6. F (Most compounds are stable relative to the elements at 25°C,
      1 atm.)
7. T (Write a balanced equation and calculate ΔG.)
8. T ($\Delta n_{gas} > 0$.)
9. T (At least mine does.)
10. F (Positive at high T.)
11. T
12. T
13. T (Entropy of gas will decrease, while that of solid will remain
      about the same; their **difference** decreases.)
14. d
15. b

16. d
17. c   (More descriptive than b.)
18. d
19. c   ($\Delta G = 0$ at equilibrium T, so $\Delta S = \Delta H/T$.)
20. a
21. c
22. b
23. c
24. d   (ΔS is positive.)
25. b
26. a
27. a   (Figure this on the basis of the signs of ΔG and ΔS.)

SELECTED READINGS

Campbell, J. A., *Why Do Chemical Reactions Occur?*, Englewood Cliffs, N. J., Prentice-Hall, 1965.
    Rather light, enjoyable reading — combines the topics of this and several other chapters (including bonding and reaction rates and mechanisms) but should be understandable.
MacWood, G. E. and F. H. Verhoek, How Can You Tell Whether a Reaction Will Occur?, *J. Chem. Ed.* (July 1961), pp. 334-337.
    A fairly concise statement of thermodynamic feasibility. (Note the use of the symbol F for Gibbs free energy.)
Porter, G., The Laws of Disorder, *Chemistry* (May 1968), pp. 23-25.
    The first of a series of articles on the energetics and dynamics of reactions. Very readable.
Sanderson, R. T., Principles of Chemical Reaction, *J. Chem. Ed.* (January 1964), pp. 13-22.
    Numerous principles are set forth (like rules) to guide the reader in predicting reaction spontaneity. Many illustrative examples are given.

Also see the readings listed in Chapter 4, particularly those of Mahan, of Pimentel, and of Strong.

13

Chemical Equilibrium in Gaseous Systems

QUESTIONS TO GUIDE YOUR STUDY

1. How would you experimentally show that a given system is in a state of equilibrium?

2. If you could watch the molecules in an equilibrium system, what would you expect to see?

3. The preceding chapter established a thermodynamic criterion for recognizing equilibrium. What is it?

4. How do you interpret equilibrium in terms of the "driving forces" behind changes in chemical systems, ΔH and ΔS?

5. How can you predict the conditions under which equilibrium may exist? For example, at what temperature and pressure are ice and water in equilibrium?

6. Can you predict the effect of changes in conditions on a system already at equilibrium? (Recall that you have been able to predict whether or not a given reaction may spontaneously occur.)

7. Is there a general approach to describing equilibrium systems? (Chapter 9 dealt with phase equilibria; chapters 16, 17, 19 and 21 deal with several types of equilibria in water solution.)

8. Can you describe a common example of a reaction that only partially converts reactants to products?

9. If many reactions do not go to completion, what does it mean about how you are to interpret chemical equations?

10. How might you experimentally show that a given reaction is reversible?

11.

12.

YOU WILL NEED TO KNOW

Concepts

1. How to interpret the signs of ΔH and ΔG – Chapters 4, 12

Math

1. How to work problems involving molar concentration units – Chapter 10

2. How to calculate ΔH and ΔG for any reaction – Chapters 4, 12

3. How to solve quadratic equations, using the quadratic formula; how to calculate square roots (e.g., by use of slide rule or logs) – Readings listed in Preface

(For the equation: $ax^2 + bx + c = 0$,

$$x = \frac{-b \pm \sqrt{b^2 - 4\,ac}}{2a}$$

Note that here only one of the two possible solutions for x will make any physical sense.)

4. How to calculate logs and antilogs – Appendix 4

CHAPTER SUMMARY

We have seen (Chapter 12) that a closed system will spontaneously undergo a change at constant temperature and pressure (i.e., a reaction will occur) if by so doing its free energy can decrease. That is, the free energy of a system moves toward a minimum value. What happens once the free energy is at this minimum value, even though, as is often the case, reactants have not yet been completely consumed? Nothing that you can see! Reaction ceases; the system, unless disturbed, is in a state of *dynamic equilibrium*. (Dynamic, because our notion that molecules are always going about their business is supported by experiment.) This state of apparent rest can be considered the result of two driving forces striking a balance: the tendency of a system to move toward a state of minimum energy; the tendency toward maximum entropy ($\Delta H - T\Delta S = 0$). Another interpretation of the equilibrium state considers that the rates of competing reactions have become equal: no net, observable change occurs as a result (Chapter 14).

The extent of a reaction for a system at equilibrium can be described by a certain ratio of concentrations. This ratio, the *equilibrium constant* K_c, is found to have a value which depends only on the temperature (much as the vapor pressure of a pure substance depends only on the temperature). Its value does not depend, for example, on the concentrations of reactants or products before equilibrium is reached, or on how the equilibrium state is approached. Of course, it does depend on the particular substances taking part in the reaction.

K_c has the same form for all equilibrium systems. For the general reaction of gaseous or dissolved species, A, B, C, and D: $aA + bB \rightleftharpoons cC + dD$

$$K_c = \frac{[C]^c \, [D]^d}{[A]^a \, [B]^b},$$

where the brackets refer to molar concentration at equilibrium. This expression is commonly referred to as the Law of Chemical Equilibrium. Like all the laws we have encountered, it is only an approximation to real behavior. The more nearly gases behave like ideal gases, the more nearly does this ratio have a constant value at a given temperature. In discussing equilibria in water solutions, we will see that the law holds very well only for very dilute solutions.

We can decide whether a given system is already in a state of equilibrium, or, if not, how it will approach equilibrium. We need only to compare the given concentrations to the ratio K_c at this same temperature. (Comparison to K_c allows us to predict the change, if any, that is spontaneous, just as we have seen whether or not the free energy could decrease.)

Also by comparison to K_c, we can often predict how a given change in conditions will affect a system initially at equilibrium. The same prediction can be made qualitatively by use of *Le Chatelier's principle:* When an equilibrium system is subjected to a change in conditions, a reaction will occur so as to minimize the effect of the change. (The system will again move toward a state of minimum free energy.)

Nothing yet can be said about how fast equilibrium is approached: neither ΔG nor K_c tell us anything about reaction rates. We will consider this topic in Chapter 14.

BASIC SKILLS

1. Given the balanced equation for a reaction involving gases, write the corresponding expression for K_c.

The principles involved here are discussed on pp. 351-3 of the text. Note particularly that terms for pure solids or liquids do not appear in the expression for K_c.

– –

What is the expression for K_c for:

$$4NH_3(g) + 7O_2(g) \rightleftharpoons 4NO_2(g) + 6H_2O(l)? \underline{\hspace{1cm}}$$

We note that the term for the gaseous product, NO_2, appears in the numerator, while those for the reactants, NH_3 and O_2, are in the denominator. No term for H_2O is included since it is a liquid. The exponents of the concentration terms are the coefficients in the balanced equation.

$$K_c = \frac{[NO_2]^4}{[NH_3]^4 \times [O_2]^7}$$

– –

See Problems 13.1 and 13.18. In certain parts of these problems, you must first write the balanced equation from the information given.

2. Given K_c for a particular equation, obtain K_c for other equations that could be written to represent the same equilibrium system.

This skill is discussed on p. 352 and is illustrated in the following example.

– –

For the equation $N_2(g) + O_2(g) \rightleftharpoons 2NO(g)$, $K_c = 0.10$. What is K_c for the equation $2NO(g) \rightleftharpoons N_2(g) + O_2(g)$? $\underline{\hspace{1cm}}$ for the equation $2N_2(g) + 2O_2(g) \rightleftharpoons 4NO(g)$? $\underline{\hspace{1cm}}$

Here it is helpful to write the several expressions for K_c.

1st expression	2nd expression
$K_c = \dfrac{[NO]^2}{[N_2] \times [O_2]} = 0.10$	$K_c' = \dfrac{[N_2] \times [O_2]}{[NO]^2} = ?$

3rd expression

$$K_c'' = \frac{[NO]^4}{[N_2]^2 \times [O_2]^2} = ?$$

Comparing the second expression to the first, it should be clear that K_c' is the reciprocal of K_c.

$$K_c' = \frac{1}{K_c} = \frac{1}{0.10} = 10$$

Comparing the third expression to the first, we see that K_c'' is the square of K_c.

$$K_c'' = (K_c)^2 = (0.10)^2 = 0.010$$

– –

This skill is applied in Problems 13.2 and 13.19. It will be useful later when we consider equilibria in aqueous solution, particularly in Chapters 17 and 18 (acid-base equilibria).

3. For a given equation, calculate the numerical value of K_c knowing:

a) the equilibrium concentrations of all species.

-- --

For the reaction $2NO(g) + O_2(g) \rightleftharpoons 2NO_2(g)$, the equilibrium concentrations of NO_2, NO, and O_2 are 1.0×10^{-2}, 1.0×10^{-3}, and 0.50×10^{-1} mole/liter respectively. What is K_c? _____

All that is required here is to set up the expression for K_c and substitute the concentrations given:

$$K_c = \frac{[NO_2]^2}{[NO]^2 \times [O_2]} = \frac{(1.0 \times 10^{-2})^2}{(1.0 \times 10^{-3})^2 \times (0.50 \times 10^{-1})} = 2.0 \times 10^3$$

-- --

This skill is applied in an extremely straightforward way in Problem 13.3; Problem 13.20 is slightly more subtle.

b) the original concentrations of all species and the equilibrium concentration of one species.

This skill is slightly more difficult to apply than 3(a). You may find it helpful to set up a table such as the one in the example below.

-- --

For the reaction $2HI(g) \rightleftharpoons H_2(g) + I_2(g)$, if we start with pure HI at a concentration of 1.20 mole/liter, that of H_2 at equilibrium is 0.20 mole/liter. What is K_c for the reaction? _____

To work this problem, you must recall the significance of the coefficients of a balanced equation: they give the mole ratios between reactants and products. Since H_2 and I_2 both have coefficients of 1, it follows that when 0.20 mole/liter of H_2 is produced, an equal number, or 0.20 mole/liter, of I_2 must be formed simultaneously. Since 2 moles of HI are required to give 1 mole of H_2, $2(0.20) = 0.40$ mole/liter of HI must be consumed to produce 0.20 mole/liter of H_2. Finally, $1.20 - 0.40 = 0.80$ mole/liter of HI must be left. Summarizing this reasoning in the form of a table:

	orig. conc. (mole/liter)	change (mole/liter)	equil. conc. (mole/liter)
HI	1.20	−0.40	0.80
H_2	0.00	+0.20	0.20
I_2	0.00	+0.20	0.20

With this information, K_c is readily calculated.

$$K_c = \frac{[H_2] \times [I_2]}{[HI]^2} = \frac{(0.20)(0.20)}{(0.80)^2} = 0.062$$

_____

This skill is applied in Problems 13.4 and 13.21 (13.4 is the easier of the two). Note that in these and many other equilibrium problems, you may be given amounts in moles rather than concentrations in moles/liter. If that is the case, it may be simplest to set up the equilibrium table in terms of numbers of moles, waiting until the last step to obtain concentrations by dividing through by the volume. However you do it, the distinction between amount and concentration is an important one, which you should have learned in Chapter 10.

4. Given the value of K_c, predict:

a) the direction in which a chemical system will move to reach equilibrium.

The principle involved here is discussed on p. 354 and illustrated in Example 13.1. Notice the importance of distinguishing between "original" or arbitrary concentrations, which may have any value, and equilibrium concentrations, which must be in a fixed ratio given by K_c. The symbol [   ] is used throughout the text for equilibrium concentrations.
This skill is applied in Problems 13.5 and 13.22.

b) the equilibrium concentration of one species, given the equilibrium concentrations of all other species.

This is a quite straightforward calculation illustrated in the following example.

_____

At 2000°C, K_c = 0.10 for the reaction $N_2(g) + O_2(g) \rightleftharpoons 2NO(g)$. If the equilibrium concentrations of N_2 and O_2 are 0.050 and 0.010 mole/liter respectively, what is the equilibrium concentration of NO? _____
Setting up the expression for K_c:

$$K_c = \frac{[NO]^2}{[N_2] \times [O_2]} = 0.10 = \frac{[NO]^2}{(0.050)(0.010)}$$

Solving: $[NO]^2 = 0.10 \times 0.050 \times 0.010 = 5.0 \times 10^{-5}$

$[NO] = (5.0 \times 10^{-5})^{1/2} = (50 \times 10^{-6})^{1/2} = 7.1 \times 10^{-3}$ mole/liter

_____

See also Problems 13.6 and 13.23

c) the equilibrium concentrations of all species, given their initial concentrations.

This is probably the most useful application of the equilibrium constant expression and the one students ordinarily find most difficult. Read carefully Example 13.2 and the comments that follow on p. 357. It may be worth cautioning against some of the mistakes commonly made in analyzing problems of this type.

- Don't confuse original and equilibrium concentrations. In part (a) of Example 13.2, you might be tempted to set $[N_2] = [O_2] = 1.0$ mole/liter, which would lead to a quite different (and incorrect!) answer. Remember, since some NO is formed, the concentrations of reactants must decrease; you can't get something for nothing!
- Be sure you relate properly the changes in reactant and product concentrations. Referring again to Example 13.2a, students occasionally take the change in NO concentration to be x rather than $2x$, forgetting that the coefficients of the balanced equation require a 2:1 mole ratio of NO to N_2.
- Don't confuse amounts with concentrations. You may note that in Example 13.2, the 10 liter volume "cancels out." Don't make the mistake of assuming that this will ordinarily happen; in general, it will not.

Calculations of this type are illustrated most simply by Problems 13.7 and 13.24. Problems 13.8, 13.10, 13.11, 13.25, 13.27, and 13.28 are similar but involve other types of calculations or assumptions as well.

d) the effect of changing the number of moles of reactant or product on the position of an equilibrium.

The direction in which the shift will occur is easily predicted from the general principle given in bold type at the bottom of p. 358. The calculations involved are illustrated in Example 13.3. Notice that the approach used is to take the concentrations just after the addition of NO_2 to be the "original" concentration. Then, a calculation essentially identical to that followed in Example 13.2a is carried out to obtain "final" equilibrium concentrations.

Problems 13.13 and 13.30 are set up in precisely the same way as Example 13.3. Unfortunately, the mathematics is somewhat more complex; the quadratic formula must be used in each case.

e) the effect of changing the volume on the position of an equilibrium.

The principle involved here is given in bold type on p. 361. The calculations are handled in much the same way as for a change in the number of moles of reactants or products.

- -

For the system $N_2O_4(g) \rightleftharpoons 2NO_2(g)$, it is found that at equilibrium 2.50 moles of N_2O_4 and 15.0 moles of NO_2 are present in a 10.0 liter container. What is the value of K_c? _____ How many moles of N_2O_4 would be present in a 2.00 liter container? _____

The equilibrium constant is readily evaluated by setting up the expression and substituting the equilibrium concentrations of both species.

$$K_c = \frac{[NO_2]^2}{[N_2O_4]} = \frac{(15.0/10.0)^2}{(2.50/10.0)} = \frac{(1.50)^2}{0.250} = 9.00$$

To answer the second question asked above, it is convenient to set up a table.

	no. moles orig.	change in no. moles	final no. moles	final conc.
N_2O_4	2.50	+x	2.50 + x	(2.50 + x)/2.00
NO_2	15.0	-2x	15.0 - 2x	(15.0 - 2x)/2.00

Notice that a decrease in volume should favor the formation of N_2O_4 so we take its change to be positive, i.e., +x. It follows from the coefficients of the balanced equation that the change in the number of moles of NO_2 must be -2x. Setting up the expression for K_c:

$$K_c = \frac{[NO_2]^2}{[N_2O_4]} = 9.00 = \frac{\left(\dfrac{15.0 - 2x}{2.00}\right)^2}{\dfrac{2.50 + x}{2.00}}$$

Without going through the gory details of the calculation, let us simply point out that the above equation can eventually be simplified to the form:

$$4x^2 - 78x + 180 = 0$$

which can be solved with the aid of the quadratic formula to give x = 2.68 moles. We refer back to the equilibrium table to find the quantity asked for:

no. moles N_2O_4 = 2.50 + 2.68 = 5.18 moles

- -

Problems 13.14 and 13.31 give you a chance to apply this skill. In both cases, the quadratic formula must be used to obtain a numerical answer.

5. Given K_c at one temperature and ΔH for the reaction, calculate K_c at another temperature, using Equation 13.6.

This skill is illustrated in Example 13.4. Note the similarity to vapor pressure calculations using the Clausius-Clapeyron equation (Chapter 9); compare, for instance, Example 9.2.

Problem 13.15 is essentially identical with Example 13.4. In Problem 13.32, the argument is turned around; you are asked to calculate the temperature at which K_c has a certain value.

*6. Predict qualitatively what effect a change in the number of moles, volume, pressure, or temperature will have on the position of an equilibrium.

The general principles may be summarized as follows:

a) When a species is added to a system at equilibrium, reaction occurs in such a direction as to consume part of the added species.

b) When the volume is decreased, reaction occurs in such a direction as to decrease the number of moles of gas. An "increase in pressure" is often taken to be equivalent to a decrease in volume. See, however, the discussion at the top of p. 362 of the text.

c) When the temperature is increased, the endothermic reaction is favored.

Notice that each of these principles can be expressed quantitatively, applying skills 4d, 4e, and 5 above. The qualitative concepts are illustrated in Problems 13.12 and 13.29.

*7. Relate the equilibrium constant for a reaction to the standard free energy change.

The basic relationship here is:

$$\Delta G^{1M} = -(2.30)\,(1.99)\,T\,\log_{10} K_c$$

or, at 298° K:

$$\Delta G^{1M} = -1360 \log_{10} K_c$$

The use of this relationship is illustrated in Example 13.5 and in Problems 13.16 and 13.33.

Problems

1. Set up the expressions for K_c for

$$P_4\,(g) + 6Cl_2\,(g) \rightleftharpoons 4PCl_3\,(g)$$

$$Al_2O3(s) + 3CO(g) \rightleftharpoons 2Al(l) + 3CO_2\,(g)$$

2. If K_c for the second reaction in (1) is 2.50, calculate K_c for:

$$1/3Al_2 O_3 \text{ (s)} + CO(g) \rightleftharpoons 2/3Al(l) + CO_2 \text{ (g)}$$

$$2Al(l) + 3CO_2 \text{ (g)} \rightleftharpoons Al_2 O_3 \text{ (s)} + 3CO(g)$$

3. Calculate the value of K_c for the reaction $2SO_3 \text{ (g)} \rightleftharpoons 2SO_2 \text{ (g)} + O_2 \text{ (g)}$ if:

a) the equilibrium concentrations of SO_2, O_2, and SO_3 are 1.6×10^{-2}, 1.2×10^{-1}, and 3.2×10^{-2} mole/liter respectively.
b) starting with pure SO_3 at a concentration of 2.6×10^{-4} mole/liter, the equilibrium concentration of SO_2 is found to be 1.4×10^{-4} mole/liter.

4. For the reaction $H_2 \text{ (g)} + Cl_2 \text{ (g)} \rightleftharpoons 2HCl(g)$, we have $K_c = 4.0$.

a) If one starts with H_2, Cl_2 and HCl at concentrations of 1.0, 0.50, and 3.0 mole/liter respectively, which way will the system move to reach equilibrium?
b) If the equilibrium concentrations of HCl and H_2 are 2.0 and 0.60 mole/liter respectively, what is $[Cl_2]$?
c) If one starts with concentrations of H_2 and Cl_2 of 1.0 mole/liter each, what is the equilibrium concentration of HCl?

4. For the system $Cl_2 \text{ (g)} \rightleftharpoons 2Cl(g)$, equilibrium is reached in a 1.0 liter container when 0.50 mole of Cl_2 and 0.40 mole of Cl are present. Check to confirm that K_c for this reaction is 0.32.

d) If one mole of Cl_2 is added to the equilibrium system, what will be the final equilibrium concentrations of Cl and Cl_2?
e) If the volume of the container is increased to 10 liters, what will be the equilibrium concentrations of Cl and Cl_2?

5. For a certain reaction, $K_c = 1.0 \times 10^{-2}$ at $27°C$ and ΔH is +23,200 cal. Calculate K_c at $127°C$.

*6. Consider the system $2N_2O(g) \rightleftharpoons 2N_2(g) + O_2(g)$, for which $\Delta H = +39.0$ kcal. In which direction will the system move to re-establish equilibrium if:

a) N_2O is added
b) O_2 is removed
c) the volume is increased
d) the pressure is increased by adding N_2O at constant volume
e) the pressure is increased by compressing the mixture
f) the pressure is increased by adding He at constant volume
g) the temperature is increased

*7. For a certain reaction, ΔG^{1M} is -13,200 cal at 298° K. Calculate K_c.

- - - - - - - - - - - -

8. For the reaction $A_2 (g) + B_2 (g) \rightleftharpoons 2AB(g)$, $K_c = 0.700$.

a) Write an expression for the equilibrium constant.
b) If the equilibrium concentrations of A_2 and B_2 are both 2.00 moles/liter, calculate $[AB]$.
c) If one starts with the concentrations of both A_2 and B_2 at 2.00 moles/liter, calculate $[AB]$.

9. For the reaction $PCl_5 (g) \rightleftharpoons PCl_3 (g) + Cl_2 (g)$, we have $K_c = 0.60$. If one starts with 1.0 mole each of PCl_5, PCl_3, and Cl_2 in a 1.0 liter container:

a) Which way will the system shift to achieve equilibrium?
b) How many moles of Cl_2 will be present at equilibrium?
c) How would your answers to (a) and (b) differ if the volume were 10.0 liters instead of 1.0 liter?
d) How would your answers to (a) and (b) differ if you started with 2.0 moles of PCl_5 instead of 1.0 mole?

10. For the equilibrium $N_2 (g) + 3H_2 (g) \rightleftharpoons 2NH_3 (g)$, it is found that when one starts with concentrations of N_2 and H_2 of 1.0 and 2.0 mole/liter, respectively, at 700° K, the equilibrium concentration of NH_3 is 0.60 mole/liter.

a) Calculate K_c at 700° K.
b) Calculate K_c at 700° K for $2NH_3 (g) \rightleftharpoons 3H_2 (g) + N_2 (g)$.
c) Calculate K_c at 700° K for $NH_3 (g) \rightleftharpoons 3/2H_2 (g) + 1/2N_2 (g)$.
d) Given that ΔHf of NH_3 is -11.0 kcal/mole, use the value of K_c calculated in (a) to obtain K_c for the same reaction at 1000° K.

- - - - - - - - - - - - -

11. At a certain temperature, $K_c = 9.0$ for the reaction

$$SO_2 (g) + NO_2 (g) \rightleftharpoons SO_3 (g) + NO(g).$$

If we start with 1.00 mole each of all four compounds in a 1.00 liter container, what should be the equilibrium concentration of NO?

12. Consider the reaction $2HI(g) \rightleftharpoons H_2 (g) + I_2 (g)$.

a) At 1000°C, 25% of a sample of HI is found to decompose to the elements at equilibirum. Calculate K_c.

b) At this same temperature, only 0.4% of any sample of HBr is decomposed to the elements at equilibrium. Account for the difference in HI and HBr behavior.

SELF-TEST

True or False

1. A chemical equation describes the relative changes in the   ( )
numbers of moles of reactants and products as equilibrium is approached.

2. Equilibrium can be considered as a balance struck between   ( )
two opposing tendencies: a system will tend to move toward a state of maximum energy; a system will tend to move toward a state of minimum entropy.

3. The pressure exerted by a sample of gaseous benzene in   ( )
equilibrium with liquid benzene will depend on the container volume.

4. Higher temperatures would favor the production of more   ( )
product in the following system: Benzene(l) $\rightleftharpoons$ Benzene(g).

5. The expression for K_c always shows all gaseous or dissolved   ( )
species, never pure solid or liquid species.

6. Gaseous hydrogen and oxygen are allowed to react to form   ( )
liquid water. The value of K_c for the reaction: $2 H_2 (g) + O_2 (g) \rightleftharpoons 2 H_2O(l)$ will depend upon the initial relative amounts of H_2 and O_2.

7. One mole of HI(g) and one mole of $H_2 (g)$ are placed in an   ( )
evacuated container at 100°C and allowed to come to equilibrium. As equilibrium is approached, one can be sure that the H_2 concentration will increase. (The reaction: $2 HI(g) \rightleftharpoons H_2(g) + I_2 (g)$)

8. The value of K_c is expected to increase with temperature for ( )
any reaction that has a negative value of ΔH.

Multiple Choice

9. The expression for K_c for the equilibrium: $C(s) + CO_2 (g) \rightleftharpoons$ ( )
$2\ CO(g)$ is

a) $\dfrac{2\ [CO]}{[C]\ [CO_2]}$

b) $\dfrac{2\ [CO]}{[CO_2]}$

c) $\dfrac{[CO]^2}{[CO_2]}$

d) $\dfrac{[CO]}{[CO_2]}$

10. Approximately stoichiometric amounts of two reactants are ( )
mixed in a suitable container. Given sufficient time, the reactants
may be converted almost entirely to products if
a) K_c is much less than one
b) K_c is much larger than one
c) the free energy change is zero
d) the free energy change is a large positive number

11. At a certain temperature, $K_c = 1$ for the reaction: ( )
$2\ HCl(g) \rightleftharpoons H_2 (g) + Cl_2 (g)$. In this system, then, one can be sure
that
a) $[HCl] = [H_2] = [Cl_2] = 1$
b) $[H_2] = [Cl_2]$
c) $[HCl] = 2 \times [H_2]$
d) $\dfrac{[H_2]\ [Cl_2]}{[HCl]^2} = 1$

12. Given the equilibrium constants for the following reactions: ( )

$$2\ Cu(s) + \tfrac{1}{2} O_2 (g) \rightleftharpoons Cu_2 O(s),\ K_1$$
$$Cu_2 O(s) + \tfrac{1}{2} O_2 (g) \rightleftharpoons 2\ CuO(s),\ K_2,$$

one can show that for:

$$2\ Cu(s) +\quad O_2 (g) \rightleftharpoons 2\ CuO(s),\ K_c \text{ is equal to}$$

a) $K_1 + K_2$

b) $K_2 - K_1$

c) $K_1 \times K_2$

d) K_2 / K_1

13. What would you predict to be the conditions that would ( )
favor maximum conversion of noxious nitric oxide and carbon
monoxide: $NO(g) + CO(g) \rightleftharpoons ½ N_2 (g) + CO_2 (g)$, $\Delta H = -89.3$ kcal?
a) low T, high P   b) high, T, high P
c) low T, low P   d) high T, low P

14. The position of equilibrium would not be affected by ( )
changes in container volume for the system:

    a) $H_2(g) + I_2(s) \rightleftharpoons 2\ HI(g)$
    b) $N_2(g) + O_2(g) \rightleftharpoons 2\ NO(g)$
    c) $N_2(g) + 3\ H_2(g) \rightleftharpoons 2\ NH_3(g)$
    d) $H_2O_2(l) \rightleftharpoons H_2O(l) + \frac{1}{2} O_2(g)$

15. For the reaction: ( )
    $4\ NH_3(g) + 7\ O_2(g) \rightleftharpoons 2\ N_2O_4(g) + 6\ H_2O(l)$,
increasing the pressure by the addition of neon gas would be
expected

    a) to increase the yield of N_2O_4 at equilibrium
    b) to decrease the yield of N_2O_4 at equilibrium
    c) to speed up the reaction in the forward direction
    d) to make no change in the relative amounts of NH_3 and
       N_2O_4 at equilibrium

16. For the reaction in (15), a decrease in pressure brought about ( )
by an increase in container volume would be expected to

    a) increase the yield of water at equilibrium
    b) diminish the extent of the forward reaction
    c) increase the value of K_c
    d) have no effect on the equilibrium attained

17. Which of the following changes will invariably increase the ( )
yield of products at equilibrium?

    a) an increase in temperature
    b) an increase in pressure
    c) addition of a catalyst
    d) increasing reactant concentrations

18. In which of the following cases will the least time be required ( )
to arrive at equilibrium?

    a) K_c is very small   b) K_c is approximately one
    c) K_c is very large   d) cannot say

19. In order to reach equilibrium in a shorter time interval, which ( )
one of the following would be appropriate to most any chemical
reaction?

    a) decrease the concentrations of reacting substances
    b) increase the temperature and pressure
    c) decrease the temperature
    d) use only stoichiometric amounts of reactants

20. For the following reaction, one would predict the form of K_c ( )
to be:

$$AgCl(s) + 2 NH_3(aq) \rightleftharpoons Ag(NH_3)_2{}^+(aq) + Cl^-(aq)$$

a) $\dfrac{[Ag(NH_3)_2{}^+] + [Cl^-]}{[NH_3]^2}$

b) $[Ag^+] [Cl^-]$

c) $\dfrac{[Ag(NH_3)_2{}^+] [Cl^-]}{[NH_3]^2}$

d) $\dfrac{1}{[AgCl]}$

21. Into a one-liter flask at $400°C$ are placed one mole of N_2, ( )
three moles of H_2 and two moles of NH_3. If K_c for the following
reaction is about 0.5 at $400°C$, what reaction, if any, can be
expected to occur?

$$N_2(g) + 3 H_2(g) \rightleftharpoons 2 NH_3(g)$$

a) left to right
c) system is at equilibrium

b) right to left
d) cannot say

22. For the reaction considered in (21), the value of K_c is about ( )
0.08 at $500°C$. One can therefore say that
a) the reaction is endothermic
b) the reaction is exothermic
c) K_c is independent of temperature
d) K_c is directly proportional to the absolute temperature

23. If we start with 1.0 mole of N_2, 3.0 moles of H_2 and 2.0 ( )
moles of NH_3 in a one-liter container at $500°C$ (where $K_c = 0.08$ for
the reaction written as in 21), at equilibrium
a) the number of moles of N_2, H_2, and NH_3 will be in the
ratio 1:3:2
b) the number of moles of N_2 and H_2 will be in the ratio
1:3
c) the number of moles of N_2 will be 1.0
d) the total number of moles will be the same as at $400°C$

24. How do the equilibrium constants for forward and reverse ( )
reactions compare to each other?
a) they are always the same
b) their sum must equal one
c) their product must equal one
d) they are not related

25. If a system contains SO_2, O_2, and SO_3 gases at equilibrium,   ( )
$$SO_2 (g) + \tfrac{1}{2} O_2 (g) \rightleftharpoons SO_3 (g)$$
the addition of more oxygen to the system will result in
    a) a reaction in which some SO_2 is formed
    b) a reaction in which part of the added O_2 is consumed
    c) a reaction in which all of the added O_2 is consumed
    d) no reaction

26. It is sometimes the case that K_c is very small for a reaction   ( )
under almost all conditions. Yet the reaction may be used to produce
significant amounts of products. How can this be?
    a) a catalyst is used to increase the yield
    b) the reaction is carried out at very high temperatures
    c) an alternate series of reactions, giving the same result, is
        used
    d) product is removed from the system as it is formed

27. For the reaction $CaO(s) + CO_2 (g) \rightleftharpoons CaCO_3 (s)$, we find that   ( )
$K_c = 277$ at 800°C. What is the equilibrium concentration of CO_2 ?
    a) 277 M   b) (1/277)M
    c) $(277)^{1/2}$M   d) not enough data given

28. Under certain conditions, the reaction in (27) gives 1.00 mole   ( )
of CO_2 in the reaction vessel with both solids present at equilibrium.
The heat of reaction (for the equation as written) is –42 kcal. Adding
more $CaCO_3 (s)$ will have what effect?
    a) increase the concentration of CO_2
    b) leave the CO_2 concentration unchanged
    c) decrease the CO_2 concentration
    d) the effect depends on the temperature

29. At a certain temperature, $K_c = 1.2 \times 10^3$ for the equilibrium   ( )
$CO(g) + Cl_2 (g) \rightleftharpoons COCl_2 (g)$. If excess O_2 were added so that any
$CO(g)$ present were removed (by conversion to CO_2 and not to
$COCl_2$), what would happen to the $COCl_2 (g)$ initially present in the
equilibrium mixture?
    a) it would vanish
    b) more would be formed
    c) its amount would be slightly diminished
    d) nothing would happen to it

30. For the reaction H_2 (g) + Br_2 (g) $\rightleftharpoons$ 2 HBr(g), K_c = 4.0 × 10^{-2}. (   )
For the reaction HBr(g) $\rightleftharpoons$ ½ H_2 (g) + ½ Br_2 (g), K_c is:
    a) 4.0 × 10^{-2} b) 2.0 × 10^{-1}
    c) 5.0   d) 25

31. For the reaction 2 SO_3 (g) $\rightleftharpoons$ 2 SO_2 (g) + O_2 (g), K_c = 32. If (   )
[SO_3] = [O_2] = 2.0 M, then [SO_2] is:
    a) 0.031   b) 0.25
    c) 5.7   d) 8.0

SELF-TEST ANSWERS

  1. T (An equation does not, for example, tell us how fast equilibrium is approached.)

  2. F (Minimum energy, maximum entropy both favor spontaneity.)

  3. F (Equilibrium vapor pressure depends only on temperature.)

  4. T

  5. T

  6. F (K_c for a given reaction depends only on temperature.)

  7. T (Hint: I_2 must form.)

  8. F

  9. c

10. b

11. d (The composition could be anything, provided this ratio equals K_c.)

12. c (Hint: Write out all three expressions for K_c.)

13. a

14. b (Numbers of moles of gas unchanged.)

15. d (Neon takes no part in the net reaction.)

16. b (Favoring the larger number of moles of gas.)

17. d

18. d (See Chapter 14: K_C, like ΔG, says nothing about reaction speed.)

19. b (Again, Chapter 14.)

20. c (Dissolved species only!)

21. a (Concentration quotient is less than K_c.)

22. b

23. b

24. c

25. b

26. d (Removal of product shifts equilibrium to form more product.)

27. b
28. b
29. a   (By decomposing to CO + Cl$_2$.)
30. c
31. d

SELECTED READINGS

Guggenheim, E. A., More about the Laws of Reaction Rates and of Equilibrium, *J. Chem. Ed.* (November 1956), pp. 544-545.
   A brief discussion of the history of the law of equilibrium and errors commonly made about the law. Follows the discussion in the article by Mysels.
Mysels, K. J., The Laws of Reaction Rates and of Equilibrium, *J. Chem. Ed.* (April 1956), pp. 178-179.
   A short discussion of errors commonly made in relating the law of equilibrium to rates of reactions. (Perhaps more appropriate reading after Chapter 14.)

Further readings, at the level of the text as well as somewhat below, can be found in the list for Chapter 12.

Further work in solving problems on chemical equilibrium can be taken from the problem manuals listed in the Preface.

14

Rates of Reaction

QUESTIONS TO GUIDE YOUR STUDY

1. Knowing what reactions *can* occur (criteria have been established in Chapters 12 and 13), can you now say what reactions *will* occur? In particular, can you say when a given reaction will begin, how fast it will go, and when it will stop?

2. What properties of a system could you observe to see how fast a reaction is occurring? What kinds of measurements would you make? How can you measure the rates for very fast and very slow reactions?

3. How does the rate of a reaction depend on conditions such as temperature, pressure and the physical state of the reactants?

4. How does the rate of a reaction depend on concentration? How do you experimentally arrive at such a relationship (the *rate law*)?

5. How do you account for a particular rate law in terms of molecular rearrangements? (The sum of these *elementary reactions* constitutes the *reaction mechanism*.) Is there ever more than one possible mechanism?

6. How do you account for the fact that most reactions speed up as temperature is increased? Can you quantitatively relate rate and temperature? (Recall that kinetic theory relates molecular speeds and temperature.)

7. How does the rate of a reaction change with time? (Recall that for closed systems, a reaction is expected to approach equilibrium.)

8. How can you change or control the rate of a reaction? (Any common examples you can think of?) How are reaction rates controlled or altered in living organisms?

9. Can you predict or calculate the rate, or write the rate law, for a particular reaction without ever carrying out that reaction? (Recall that you have been able to calculate ΔH, ΔG, K_c for such a case.)

10. In what ways are rate laws useful? What, if anything, do they let you say about the feasibility of carrying out a particular reaction?

11.

12.

YOU WILL NEED TO KNOW

Concepts

1. How to interpret the sign of ΔH — Chapter 4
2. How to interpret the Maxwell distribution of molecular energies — Chapter 5

Math

1. How to calculate logs and antilogs — Appendix 4
2. How to work problems involving molar concentration units — Chapter 10
3. How to work with the graph of an equation for a straight line (e.g., $\log k = -B/T + A$ has the form $y = ax + b$) — See a math text or Preface.

CHAPTER SUMMARY

From the free energy change (Chapter 12) or the equilibrium constant (Chapter 13), we can determine whether or not a reaction will take place and, if so, the extent to which it will occur, *given sufficient time*. Thermodynamic quantities such as ΔG and K_c are derived from experimental measurements; they require no knowledge of molecular behavior. In contrast, when we deal with rate of reaction we try to establish the path or mechanism by which it occurs. This requires that we make some assumptions as to what the molecules are up to. Hopefully, these assumptions can be checked by experimental measurements which tell us how reactant concentrations change with time.

The simplest molecular model for reaction path assumes that in order for reaction to occur, two high energy molecules must collide. This implies that reaction rate should increase with concentration: the more molecules there are in a given volume, the more collisions there will be in unit time. Experimentally, we find that rate does indeed increase with concentration. However, the relationship is more complicated than a simple collision mechanism would imply. If the path of every reaction involved nothing more than a simple, two-molecule collision, all reactions would be second order.

That is, in the general rate expression

$$A + B \rightarrow \text{products}; \text{rate} = k(\text{conc A})^m (\text{conc B})^n$$

we would expect that: $m + n = 2$. Experimentally, we find that reactions of other orders are common. In particular, we frequently encounter first order reactions where the relationship between concentration and time is given by the equation:

$$\log_{10} \frac{X_0}{X} = \frac{kt}{2.30}$$

where X_0 and X are reactant concentrations at times $t = 0$ and t respectively. Reactions of other integral orders (0, or 3) or fractional orders ($\frac{1}{2}, \frac{3}{2}$) are also known.

The basic weakness of the simple collision model is that very few reactions proceed by a single step. More frequently, the reaction mechanism involves a series of steps, each of which may be a bimolecular collision or the decomposition of an unstable, high-energy species. In most cases, one step is considerably slower than the others and hence determines both the overall rate and the observed reaction order. Experimental data on reaction rate may suggest what this step is but it can never (?) establish an unambiguous mechanism for a particular reaction.

The collision theory explains quite satisfactorily the dependence of reaction rate upon temperature. If we assume that only collisions between high-energy molecules are effective, we can derive the Arrhenius relation between rate constant and temperature:

$$\log_{10} k = \text{constant} - \frac{E_a}{(2.30) \, RT}$$

The activation energy, E_a, represents the minimum energy that must be available if a collision is to be fruitful. We see from the Arrhenius equation that a reaction which requires a high activation energy will ordinarily be slow (small rate constant). Conversely, if E_a is very small, most of the molecules have sufficient energy to react when they collide and reaction should occur very rapidly.

It may be possible to lower the activation energy by finding a different reaction path with a lower energy barrier. In the laboratory, we do this by finding an appropriate catalyst for the reaction. Frequently, a catalyst provides an active surface upon which reaction can occur. Remember that a catalyst changes only the activation energy and hence the rate of reaction; it does not affect the overall energy difference between products and reactants and hence cannot change the equilibrium constant.

BASIC SKILLS

1. Determine the order of a reaction, given the initial rate as a function of concentration of reactants.

This skill is illustrated in Example 14.1. Note (in parts b and c) that data of this type can also be used to obtain the rate constant for a reaction and to predict rates at any given concentration.

Problems 14.4 and 14.19 are entirely analogous to Example 14.1. Problems 14.3 and 14.18 involve a similar approach, except that the reasoning is turned around; assuming a particular order, you are asked to calculate a rate constant (14.18) or a rate at a given concentration (14.3).

2. Use the rate equation for a first order reaction (Equation 14.5) to obtain:

a) the concentration of reactant after a given time, knowing its original concentration and the rate constant;
b) the time required for the concentration of reactant to drop to a particular value, given the rate constant and the original concentration.

The use of the first order rate equation for these purposes is illustrated in Example 14.2 parts (a) and (b). Notice that there is some advantage in retaining the quantity "log X_0/X" as a unit in your calculations. Since X_0 is always greater than X, log X_0/X is always a positive quantity. Most sutdents prefer to avoid using negative logarithms whenever possible.

Problems 14.6a and 14.21 are straightforward examples of calculations involving concentration-time relations in first order reactions. Problems 14.7 and 14.22 are slightly more difficult but really involve little more than substituting into the first order rate equation.

3. Given either the half-life or the rate constant for a first order reaction, calculate the other quantity.

For a first order reaction, the relation between these two quantities is: $t_{1/2} = 0.693/k$. This relation is derived and then applied in Example 14.2, part (c), p. 378. See also Problem 14.6b.

4. Use the Arrhenius equation (Equation 14.9) to obtain

a) the rate constant at T_2, given its value at T_1 and the activation energy;
b) the activation energy, given rate constants at two different temperatures;

c) the temperature at which k will have a specified value, given E_a and k_1 at T_1.

Skills 3a and 3b are illustrated in Example 14.4, parts (a) and (b). The example below applies the same reasoning for 3c.

The activation energy for a certain reaction is 6400 cal. The rate constant is 1.00×10^{-2}/sec at $27°$C. At what temperature will k be 2.00×10^{-2}/sec?

Looking at Equation 14.9, we see that we are given all the quantities except T_2.

$$k_1 = 1.00 \times 10^{-2}/\text{sec}; T_1 = (27 + 273)°K = 300°K$$

$$k_2 = 2.00 \times 10^{-2}/\text{sec} \qquad E_a = 6400 \text{ cal}$$

Substituting:

$$\log_{10} \frac{2.00 \times 10^{-2}}{1.00 \times 10^{-2}} = \log_{10} 2.00 = \frac{6400 (T_2 - 300)}{(2.30) (1.99) (300)T_2}$$

Noting that $\log_{10} 2.00 = 0.301$ and that $6400/(2.30) (1.99) (300) = 4.66$, we have:

$$0.301 = \frac{4.66(T_2 - 300)}{T_2}$$

Solving: $\quad 0.301 T_2 = 4.66 T_2 - 4.66(300); T_2 = 321°K$ or $48°C$.

Note the similarity between Equation 14.9, which describes how the rate constant varies with temperature, Equation 13.6, p. 363, which tells us how K_c changes with temperature, and Equation 9.4, p. 240, for the variation of vapor pressure with temperature. Indeed, Examples 14.4, p. 382, 13.4, p. 363, and 9.2, p. 240 are virtually identical except for differences in wording.

Of the problems at the end of Chapter 14 that deal with activation energies, rate constants, and temperatures, Problem 14.10 is the simplest. Problems 14.9, 14.24, and 14.25 are slightly more difficult but much more interesting.

5. Determine whether a proposed mechanism for a reaction is consistent with the observed rate expression.

The general principle here is that the form of the rate expression is determined by the mechanism of the slowest step. This step often involves a very reactive intermediate which is not one of the original reactants. Such is the case with the Cl atom in Example 14.5. The rate expression observed in the laboratory cannot involve such intermediates; in the example quoted, the rate is expressed in terms of the concentrations of the two main reactants, Cl_2 and $CHCl_3$.

This skill is applied in Problems 14.13, 14.14, 14.28, and 14.29. Note that the equations written in these problems represent actual mechanisms. Thus, in Problem 14.13a, the assumption is that the reaction occurs through a simultaneous collision of two NO molecules with an O_2 molecule.

6. Given two of the three quantities, ΔH for a reaction, E_a for a reaction and E_a' for the reverse reaction, calculate the third quantity.

The appropriate relation (p. 384) is:

$$\Delta H = E_a - E_a$$

See Problems 14.11 and 14.26; in both cases, ΔH must be calculated first.

7. Use the information given in Table 14.4 to determine the order of a reaction, given experimental rate data.

The technique involved is illustrated in Example 14.3 and in Problem 14.23. One can also use the relationship between half life, $t_{1/2}$, and original concentration, X_0, to decide whether a reaction is 0, 1st, or 2nd order (see Problem 14.8).

Problems

1. For a reaction between three species, A, B, and C, the following rate data were obtained:

rate (mole/liter min)	0.010	0.020	0.040	0.010
conc. A (mole/liter)	0.10	0.20	0.10	0.10
conc. B (mole/liter)	0.10	0.10	0.20	0.10
conc. C (mole/liter)	0.10	0.10	0.10	0.20

Determine the order of the reaction with respect to A, B, and C.

2. The rate constant for a first order reaction is 2.2×10^{-2} /sec.

   a) If one starts with a reactant concentration of 0.10 mole/liter, what will be the concentration after 30 seconds?
   b) How long will it take for the concentration to drop from 1.00 mole/liter to 0.050 mole/liter?

3. What is the half-life of the reaction in (2)?

4. a) The rate constant for a certain reaction is 3.0×10^{-3}/min at 25°C. If the activation energy is 26.2 kcal, what will the rate constant be at 40°C?

b) What is the activation energy for a reaction if the rate constant doubles when the temperature increases from 25°C to 40°C?

c) For the reaction referred to in (a), at what temperature will $k = 3.0 \times 10^{-2}$/min?

5. The observed rate law for a reaction between A_2 and B_2 is:

$$\text{rate} = k(\text{conc. } A_2) \, (\text{conc. } B_2)^{1/2}$$

Which of the following mechanisms is consistent with this rate law?

a) $A_2 + B_2 \rightarrow$ products
b) $B_2 \rightleftharpoons 2B$
   $B + A_2 \rightarrow$ products; (slow)
c) $A_2 \rightleftharpoons 2A$
   $A + B_2 \rightarrow$ products; (slow)

*6. *The activation energy for a certain reaction is 21.0 kcal. ΔH for the reaction is +15.2 kcal. What is the activation energy for the reverse reaction?*

*7. *The following data are obtained for a reaction involving A:*

conc. A	1.00	0.50	0.33	0.25
t	0	10	20	30

Using the information in Table 14.4, decide by graphing whether the order of the reaction is zero, one, or two.

- - - - - - - - - - - -

8. The rate constant for a certain first order reaction is 4.6×10^{-3}/min at 27°C and 1.5×10^{-2}/min at 47°C.

a) What is the activation energy for the reaction?
b) Estimate the value of the rate constant at 100°C.
c) At what temperature will the rate constant be 3.0×10^{-2}/min?
d) What is the half-life at 27°C? at 47°C?
e) If one starts with a concentration of 0.50 mole/liter, what will be the concentration after one hour at 27°C?
f) At 47°C, how long will it take for the concentration to drop to 1/3 of its original value?

9. Given the following information for a reaction involving A and B:

rate (mole/liter min)   0.0250   0.100   0.0100
conc. A (mole/liter)   0.500   0.500   0.200
conc. B (mole/liter)   0.200   0.400   0.200

a) What is the order of the reaction with respect to A? with respect to B?
b) What is the rate constant for the reaction?
c) What is the rate when the concentrations of both A and B are 1.00 mole/liter?
d) Which of the following mechanisms are consistent with the rate expression?
   i) $A + B \rightarrow$ products
   ii) $A + B \rightleftharpoons AB$
       $AB + B \rightarrow$ products; (slow)
   iii) $A + B \rightleftharpoons AB$
       $AB + A \rightarrow$ products; (slow)

SELF-TEST

True or False

1. The rate of reaction ordinarily decreases with time.   (   )

2. If the rate of reaction doubles when the concentration is doubled, the reaction must be first order.   (   )

3. For a zero order reaction the rate constant k could have the units sec^{-1}.   (   )

4. In a first order reaction a plot of concentration vs time is a straight line.   (   )

5. In general, very fast reactions have small activation energies.   (   )

6. The activation energy for a reaction can be obtained by taking the difference in energy between reactants and products.   (   )

7. An enzyme-catalyzed reaction may be inhibited by adding a substance with a structure very similar to that of the substrate.   (   )

8. Any reaction for which the rate determining step involves a collision between two molecules must be second order.   (   )

9. The order of reactions occurring at solid surfaces frequently changes with concentration.   (   )

10. It is possible for the activation energy, as calculated from the ( )
Arrhenius equation, to be negative.

11. The reaction: $H_2(g) + I_2(g) \rightarrow 2\ HI(g)$ is first order in both ( )
H_2 and I_2. Consequently, the mechanism must involve a simple
bimolecular collision between H_2 and I_2 molecules.

Multiple Choice

12. For the reaction: $2\ NO(g) + O_2(g) \rightarrow 2\ NO_2(g)$, the rate is ( )
expressed as $-\Delta conc\ O_2/\Delta t$. An equivalent expression would be
   a) $\Delta conc\ NO_2/\Delta t$ b) $-\Delta conc\ NO_2/\Delta t$
   c) $-2\Delta conc\ NO_2/\Delta t$ d) none of these

13. For a certain decomposition the rate is 0.30 mole/lit sec ( )
when the concentration of reactant is 0.20 M. If the reaction is
second order, the rate (mole/lit sec) when the concentration is 0.60
M will be:
   a) 0.30   b) 0.60   c) 0.90   d) 2.7

14. In a first order reaction the half life is 20 minutes. The rate ( )
constant k in min^{-1} is about
   a) 0.035   b) 0.35   c) 13.9   d) cannot tell

15. For a reaction of 3/2 order, it takes 20 minutes for the ( )
concentration to drop from 1.0M to 0.60 M. The time required for
the concentration to drop from 0.60 M to 0.20 M will be
   a) more than 20 min   b) 20 min
   c) less than 20 min   d) cannot tell

16. Two reactions, R_1 and R_2, have activation energies of 40 ( )
kcal and 20 kcal, respectively. Which one of the following statements
must be true?
   a) R_1 is faster than R_2 at any given T.
   b) R_1 is slower than R_2 at any given T.
   c) the rate of R_1 increases with T more rapidly than that of
      R_2.
   d) the rate of R_1 is doubled by increasing the temperature
      $10°C$.

17. The activation energy of a certain reaction is 15 kcal. The ( )
activation energy for the reverse reaction is
   a) -15 kcal   b) >15 kcal   c) <15 kcal   d) cannot tell

18. The effectiveness of a catalyst depends upon its ability to ( )
   a) decrease the activation energy   b) increase K_c
   c) increase reactant concentration   d) increase temperature

19. The principal reason for the increase in reaction rate with ( )
temperature is:
   a) molecules collide more frequently at high temperatures
   b) the pressure exerted by reactant molecules increases with T
   c) the activation energy increases with T
   d) the fraction of high energy molecules increases with T

20. For the chain reaction between H_2 and F_2, the step: ( )
$H + F \rightarrow HF$ represents
   a) chain initiation
   b) chain propagation
   c) chain termination
   d) the overall mechanism of the reaction

21. The decomposition of ozone is believed to occur by the ( )
mechanism

$$O_3 \rightleftharpoons O_2 + O$$
$$O + O_3 \rightarrow 2 O_2 \quad \text{(rate determining)}$$

When the concentration of O_2 is increased, the rate will
   a) increase   b) decrease   c) stay the same   d) cannot say

22. Enzyme-catalyzed reactions resemble surface reactions most ( )
closely in
   a) mechanism   b) E_a c) ΔG d) ΔH

23. In the reaction: $A \rightarrow B \rightarrow C$, the concentration of the inter- ( )
mediate B is likely to
   a) increase steadily with time
   b) decrease steadily with time
   c) be independent of time throughout the reaction
   d) remain constant through most of the reaction

24. Which of the following statements is true for all zero order ( )
reactions?
   a) the activation energy is very low
   b) the concentration of reactant does not change with time
   c) the rate constant, k, is zero
   d) the rate is independent of time

25. The following mechanism is proposed for the oxidation of ( )
iodide ion

$$NO + \tfrac{1}{2} O_2 \rightarrow NO_2$$
$$NO_2 + 2 I^- + 2 H^+ \rightarrow NO + I_2 + H_2O$$
$$I_2 + I^- \rightarrow I_3^-$$

A catalyst in this reaction is

  a) NO b) I⁻ c) O_2 d) H⁺

SELF-TEST ANSWERS

1. **T** (Rate slows as reactants disappear.)
2. **T**
3. **F** (Mole/lit sec.)
4. **F**
5. **T**
6. **F**
7. **T**
8. **F** (Only if the molecules that collide are the two reactants.)
9. **T**
10. **T** (Some reactions slow down as T rises.)
11. **F** (Cannot deduce mechanism unambiguously from rate law.)
12. **d** ($\tfrac{1}{2}\Delta$conc NO_2/ΔT or, $-\tfrac{1}{2}\Delta$conc NO/ΔT.)
13. **d**
14. **a**
15. **a**
16. **c** (As T changes, compare the fraction of molecules with energy $> E_1$ to the fraction of molecules with energy $> E_2$, taking $E_1 > E_2$.)
17. **d**
18. **a**
19. **d** (The increase is exponential.)
20. **c**
21. **b** (Conc O inversely related to that of O_2 by the rapidly established equilibrium.)
22. **a**
23. **d** (Has to be zero at beginning and end.)
24. **d**
25. **a** (No net change in the NO.)

SELECTED READINGS

Bunting, R. K., Periodicity in Chemical Systems, *Chemistry* (April 1972), pp. 18-20.
   The title refers to periodic recurrences during chemical reaction in certain systems of coupled reactions.

Edwards, J. O., From Stoichiometry and Rate Laws to Mechanism, *J. Chem. Ed.* (June 1968), pp. 381-385.
   How does the chemist decide on a mechanism? This article presents some of the "working rules" followed. Somewhat advanced.

King, E. L., *How Chemical Reactions Occur: An Introduction to Chemical Kinetics and Reaction Mechanisms*, New York, W. A. Benjamin, 1964.
   A good introduction to many aspects of kinetics — catalysis, very fast reactions, experimental methods — includes exercises.

Tamaru, K., New Catalysts for Old Reactions, *American Scientist* (July-August 1972), pp. 474-479.
   Partly a general review, this article presents a discussion of some new catalysts based on graphite, with unusual properties.

Wolfgang, R., Chemical Accelerators, *Scientific American* (October 1968), pp. 44-52.
   Beams of molecules passing through a vacuum find use in the new tool for studying reaction mechanisms and energies.

   For further discussion of kinetics, and its connection with equilibrium, see the book by Campbell listed in Chapter 12.

15

The Atmosphere

QUESTIONS TO GUIDE YOUR STUDY

1. What are some of the bulk properties of the atmosphere? What are its dimensions and mass? Can the ideal gas law be applied to relate some of these properties?

2. What is the overall composition of the atmosphere? How does the composition vary with weather, geographical location and altitude? What kinds of experiments give this information; what kinds of explanations do we give?

3. How may the components of the atmosphere be separated and identified?

4. What are the origins of the atmospheric components which are most abundant? (For example: What reactions are known to give rise to atmospheric oxygen; to deplete it?)

5. What reactions do the components of the atmosphere take part in, within the atmosphere itself, as well as at the interface of atmosphere and earth and between the atmosphere and living organisms?

6. What are the rates and extents of these reactions?

7. What substances can be considered as atmospheric contaminants? Where do they come from; where do they go? What effects do they have on atmospheric properties; on life and other processes?

8. How do you test for pollutants? How do you specify their concentrations? How do you control them? How do you prevent them?

9. How do you measure the extent of pollution and its change with time? (Is pollution increasing?)

10. What are some of the reactions by which pollutants can be removed? What are their limitations; their costs?

11.

12.

YOU WILL NEED TO KNOW

Concepts

1. How to interpret the sign and relative magnitudes of thermodynamic quantities (ΔH, ΔS, ΔG) and relate these to changes at the atomic-molecular level — Chapters 4, 12

Math

1. How to predict, qualitatively as well as quantitatively, the effects of changes in reaction conditions (T, P, concentration, nature of reactant, etc.) on:

    spontaneity of reaction (sign of ΔG) — Chapter 12
    equilibrium concentrations — Chapter 13
    rate of reaction — Chapter 14

2. How to work with concentration units introduced thus far — Chapter 10

3. How to define as well as work with partial pressure — Chapter 5

CHAPTER SUMMARY

In this chapter, we have attempted to tie together the concepts of chemical kinetics, chemical equilibrium, and chemical bonding introduced in previous chapters, to discuss the chemical and physical properties of the atmosphere. Interwoven with these concepts is a considerable amount of descriptive chemistry of the major constituents of the atmosphere: nitrogen, oxygen, carbon dioxide, water vapor, and the noble gases.

Our interest in the chemistry of the very unreactive element nitrogen centers upon the process of nitrogen fixation in which N_2 is converted to useful compounds. This process is carried out in nature by certain clever bacteria found in the roots of clover and other legumes. Industrially, it is accomplished by reacting nitrogen with hydrogen to form ammonia (Haber process). Ammonia, either directly or in the form of its salts (e.g., NH_4NO_3), is used as a fertilizer. It can be converted by the Ostwald process to nitric acid, which in turn is used to make fertilizers (inorganic NO_3^- salts) and explosives (organic nitro compounds).

Oxygen reacts with all but a very few elements. With most metals, the O_2 molecule is converted to the oxide ion, O^{2-}. Certain of the 1A and 2A metals normally yield peroxides (O_2^{2-}) or superoxides (O_2^-). Perhaps the most important of the reactions of oxygen with the nonmetals is that with sulfur; the sulfur dioxide produced by combustion under ordinary conditions is converted first to sulfur trioxide and then to sulfuric acid. When an element forms more than one compound with oxygen, it is ordinarily the "higher" oxide (e.g., CuO, CO_2) which is formed at low temperatures in the presence of excess oxygen ("oxidizing atmosphere"). High temperatures and limited amounts of oxygen, a "reducing atmosphere," favor the lower oxide (e.g., Cu_2O, CO).

In the upper atmosphere, we find many species, such as O_3, O, O_2^+, which are unstable under ordinary laboratory conditions. These are formed by high energy radiation in the ultraviolet and far UV regions of the spectrum; very little of this radiation reaches the surface of the earth. Since the reactions that form these species are nonspontaneous from a thermodynamic point of view, we might expect them to be readily reversed. However, because the concentrations of these atoms and ions are extremely low, their rate of recombination to form stable molecules such as O_2 is quite slow.

Among the "minor" but objectionable constituents of the atmosphere are the oxides of sulfur (mostly SO_2), the oxides of nitrogen (mostly N_2O), carbon monoxides, hydrocarbons, and suspended particles. Even though these species are present at a level of a few parts per million or less, they pose serious problems to human health. The automobile is a major source of three of these five pollutants (NO_x, CO, hydrocarbons) and is largely responsible for smog formation, which involves reaction of NO_2 with hydrocarbons. The combustion of fuels in power plants and industry is responsible for much of the sulfur oxides and suspended solids that enter the atmosphere.

Efforts to reduce the level of air pollution have concentrated largely upon modifications of the internal combustion engine and exhaust system of the automobile. Several different approaches are under study, but none, at least at present, seem likely to reduce NO_x and hydrocarbon emissions to the level required to prevent smog formation. Perhaps the ultimate answer will be to use an energy source other than gasoline; possibilities include natural gas, electricity, and steam.

BASIC SKILLS

Very few, if any, new concepts are introduced in this chapter. If you examine the problems, you will find that most of them fall into one of three categories.

a) *The application of concepts introduced in previous chapters,* particularly Chapters 4 and 12 (Thermodynamics), 13 (Chemical Equilibrium), and 14 (Chemical Kinetics). For example, Problems 15.9 and 15.29 are similar to problems in Chapter 12; Problems 15.10 and 15.30 could equally well have appeared in Chapter 13; Problems 15.20 and 15.40 are straightforward applications of the principles of kinetics (Chapter 14).

b) *Descriptive chemistry, involving preparations or equation writing.* Problems 15.6, 15.7, 15.26, and 15.27 fall into this category. You should be able to work them with little difficulty if you have assimilated the descriptive material in this chapter, particularly in Section 15.3.

c) *Discussion-type questions* based on qualitative ideas introduced in this chapter. See Problems 15.3, 15.11, 15.13, 15.15, 15.17 to 15.19, and the corresponding answered problems in the right hand columns.

The three "skills" listed below are relatively simple; moreover, the first two have been at least partially developed in previous chapters.

1. Using conversion factors given in Table 15.2, convert concentrations of gaseous species from one unit to another.

This skill is illustrated in Example 15.1, which covers virtually every conversion of this type. Problems 15.2 and 15.22 give you a chance to practice this skill. Incidentally, it is not necessary to memorize the conversion factors given in Table 15.2. All of them follow logically, either from the definitions of the various terms or from the gas laws (Chapter 5). The only one that might puzzle you is the factor relating concentration in moles/liter to partial pressures. Recall that the Ideal Gas Law states that $PV = nRT$. Hence:

$$\frac{n}{V} = \frac{moles}{liter} = \frac{P}{RT} = \frac{P}{(0.0821)T} \quad (P \text{ in atmospheres, } T \text{ in } °K)$$

2. Calculate the wavelength of radiation that is capable of bringing about a photochemical reaction.

The fundamental relation here is the Einstein relation, introduced in Chapter 6: $E = hc/\lambda$. As used here, E is the energy which must be absorbed to bring about the reaction. It is usually given in kcal/mole. To find λ (in cm), taking $h = 6.62 \times 10^{-27}$ erg sec and $c = 3.00 \times 10^{10}$ cm/sec, E must be converted to ergs/photon, using the conversion factor:

$$1 \times 10^{-12} \text{ ergs/photon} = 14.4 \text{ kcal/mole}$$

This skill is illustrated in Example 15.3 and is applied to solve Problems 15.16 and 15.36. (Note that "visible light" has a wavelength range from 4000 to 7000 Å.)

3. Given two of the three quantities relative humidity, partial pressure of water vapor in the air, and equilibrium vapor pressure of water, calculate the third quantity.

Calculations of this type follow directly from the definition of relative humidity (Equation 15.26). See Example 15.2 and Problems 15.12 and 15.32.

Problems

1. Repeat the calculations of Example 15.1a, p. 399, for O_2 instead of N_2.

2. What is the longest wavelength of radiation that can bring about a reaction which absorbs 110 kcal/mole?

3. What is the relative humidity on a day when the partial pressure of water vapor in the air is 10.6 mm Hg and the temperature is 24°C? (Refer to Appendix 1 for a table of equilibrium vapor pressures of water.)

— — — — — — — — — — — — —

4. Suppose the partial pressure of water vapor in the air is 12.0 mm Hg when the temperature is 24°C and the total pressure is 750 mm Hg.

    a) What is the relative humidity?
    b) What is the mole fraction of water in the air? (Table 15.2)
    c) What is the concentration of water in the air in moles/liter?
    d) Based upon your answer to (b), calculate the volume % of water in the air; the weight %; the concentration in ppm; the concentration in ppb.

5. Consider light at the violet edge of the visible range (4000 Å).

    a) What is the energy of this light in ergs/photon?
    b) Will a photon of this energy be capable of breaking a bond for which E = 10 kcal/mole? 100 kcal/mole?

SELF-TEST

True or False

1. Helium is the most abundant noble gas in the atmosphere. ( )

2. Of all the gases in the atmosphere, nitrogen is the least ( )
reactive.

3. The reaction of oxygen with a 1A metal may produce O^{2-}, ( )
O_2^{2-} or O_2^- ions.

4. Carbon dioxide is obtained commercially by fractional ( )
distillation of air.

5. When a sample of air is allowed to warm up without changing ( )
the total water content, the relative humidity increases.

6. As one moves up in the atmosphere, the ratio (conc O)/(conc ( )
O_2) increases.

7. The temperature profile of the atmosphere given in Figure ( )
15.5 could, at least in principle, be obtained by taking readings on a
mercury-in-glass thermometer.

8. The ratio of O_3 to O concentration in the upper atmosphere ( )
can be calculated knowing the concentration of O_2 and the equilibrium constant for the reaction: $O_3 \rightleftharpoons O_2 + O$.

9. In order to reduce the concentration of NO in automobile ( )
exhaust, it is desirable to increase the temperature at which the fuel
is burned.

10. The equilibrium constant for the reaction: ( )

$$CO(g) + Hem.O_2(aq) \rightleftharpoons O_2(g) + Hem.CO(aq)$$

is less than one.

Multiple Choice

11. In a synthetic atmosphere made up of He and O_2, the mole ( )
fraction of He is 0.80. The weight % of He is:
    a) 80 b) less than 80 c) greater than 80 d) cannot tell

12. The formula of calcium nitride is ( )
    a) CaN   b) Ca_2N c) Ca_2N_3 d) Ca_3N_2

13. In the compound Fe_3O_4, the mole ratio of Fe^{2+} to Fe^{3+} is ( )
    a) 1:1   b) 1:2   c) 2:1   d) 3:4

14. In the Haber process for making ammonia, high pressures are   (   )
used to
    a) increase the yield, leaving the rate unchanged
    b) increase the rate, leaving the yield unchanged
    c) increase the yield and the rate
    d) increase the equilibrium constant and the rate

15. Consider the reaction: $SnO(s) + \frac{1}{2} O_2(g) \rightarrow SnO_2(s)$. This   (   )
reaction would be expected to
    a) become less spontaneous at high temperatures
    b) become less spontaneous at high pressures
    c) become less spontaneous at low temperatures
    d) not occur at any temperature or pressure

16. To increase the yield of SO_3 from SO_2 in the reaction with   (   )
O_2, we could
    a) increase the temperature
    b) use more SO_2
    c) add a catalyst
    d) increase the pressure

17. Of the following noble gases, which reacts most readily with   (   )
fluorine?
    a) He   b) Ne   c) Kr   d) Xe

18. Dry Ice is effective in seeding clouds because   (   )
    a) CO_2 and H_2O have similar crystal structures
    b) it increases the water content of the cloud
    c) CO_2 molecules offer a nucleus for condensation
    d) upon sublimation, it lowers the temperature of the water

19. Automobile emissions are not a major source of   (   )
    a) NO_2 b) CO   c) hydrocarbons   d) SO_2

20. Which one of the following hydrocarbons would be most   (   )
likely to contribute directly to smog formation?
    a) CH_4 ·   b) C_2H_4 c) C_3H_8 d) C_6H_6

21. Of the following fuels, which one, under normal conditions,   (   )
produces the lowest concentration of pollutants?
    a) coal   b) wood   c) natural gas   d) petroleum

22. The superoxide ion, found in KO_2, has the structure: (   )

    a) :O–O: b) :O=O: c) :O–O: d) :O–O:

23. Of the species: O_2, $O_2{}^{2-}$, $O_2{}^-$, O^{2-}

   a) which are paramagnetic?_____

   b) which contain covalent bonds?_____

   c) which react with water to form H_2O_2?_____

   d) which react with water to form O_3?_____

   e) which one is most stable to thermal decomposition?_____

24. Of the species: O_2, N_2, O, N, N^+

   a) which is most abundant in the atmosphere?_____

   b) which are unstable under ordinary laboratory conditions?_____

   c) which is most difficult to form from an energy stand-point?_____

   d) which are held together by covalent bonds?_____

25. Of the molecules: NO_2, SO_2, O_3, CO, CO_2

   a) which are major contributors to photochemical smog?_____

   b) which can be converted to strong acids?_____

   c) which one is most abundant in automobile exhaust?_____

   d) which ones are thermodynamically unstable in the presence of O_2?_____

SELF-TEST ANSWERS

  1. F

  2. F (Noble gases.)

  3. T

  4. F

  5. F (More vapor could be present at higher T, since vp increases with T.)

  6. T

  7. F (High T at high altitudes is calculated from molecular speeds.)

  8. F (Molecules too far apart, react too slowly to establish equilibrium.)

  9. F (Its formation becomes spontaneous at high T.)

10. F (If it were, CO would not be poisonous.)

11. b

12. d
13. b $(1 \ Fe^{2+}, 2 \ Fe^{3+} \ to \ 4 \ O^{2-})$
14. c
15. a   (Consider effect of T for $\Delta S < 0$: $\Delta G = \Delta H - T \Delta S$.)
16. d   (Is b as good an answer?)
17. d
18. d
19. d
20. b   (Reactive double bond.)
21. c
22. a
23. a) O_2, O_2^- b) O_2, O_2^{2-}, O_2^- c) O_2^{2-}, O_2^- d) none e) O^{2-}
24. a) N_2 b) O, N, N^+ c) N^+ d) O_2, N_2
25. a) NO_2, O_3 b) NO_2, SO_2 c) CO_2 d) NO_2, SO_2, O_3, CO

SELECTED READINGS

The Biosphere, San Francisco, W. H. Freeman, 1970.
    The September 1970 issue of Scientific American — *with relevant materials on the substances entering and leaving the atmosphere in "cycles." For a systems approach to "all" chemical cycles, see the article by R. Siever in the June 1974 issue.*
Kellogg, W. W., and others, The Sulfur Cycle, *Science* (February 11, 1972), pp. 587-595. 587-595.
    Outlines what is known, and what we need to know, about man's and nature's contribution to the sulfur compounds in the atmosphere and the oceans; with some forecasts for the future.
Kenyon, D. H. and G. Steinman, *Biochemical Predestination,* New York, McGraw-Hill, 1969.
    Chapter 3 discusses evidence for the current view of the composition and origin of the primitive earth atmosphere — as part of a discussion of the chemical origins of life.
Lewis, J. S., The Atmosphere, Clouds and Surface of Venus, *American Scientist* (September-October 1971), pp. 557-566.
    Couples some basic chemistry with descriptive astronomy, with an emphasis on the comparison of earth chemistry and likely (and known) chemistry of Venus.
Stoker, H. S. and S. L. Seager, *Environmental Chemistry: Air and Water Pollution,* Glenview, Ill., Scott, Foresman, 1972.
    One of the best sources for quantitative material on pollution. A lot of chemistry discussed in a straightforward manner.
Turner, B. E., Interstellar Molecules, *Scientific American* (March 1973), pp. 51-69.
    On the interesting molecules that exist beyond our atmosphere, how they are observed, how they are formed and consumed.
Chemical and Engineering News (October 2, 1967), Special Report: Chemistry and the Solid Earth.
    A very broad survey of where we stand (1967) in understanding the chemistry and physics of the geosphere. (Also see the article by Bachmann referred to in Chapter 1 of the guide.)

16

Precipitation Reactions

QUESTIONS TO GUIDE YOUR STUDY

1. What occurs during a precipitation reaction? What would you observe during such a reaction? (Can you give any common example?)

2. Why does a precipitation occur? What are the driving forces behind the reaction? What are the ions doing?

3. What substances participate in this class of reaction? Can some generalizations be made? Are there correlations with electronic structure, and hence with the periodic table?

4. How do reaction conditions such as temperature and concentration affect the spontaneity and extent of a precipitation reaction?

5. Can you apply Le Chatelier's principle to determine the effect of changes in reaction conditions? Can such predictions be made quantitative?

6. Can you predict the direction and extent for this kind of reaction? (How have you predicted spontaneity and extent, qualitatively and quantitatively, for other systems?)

7. How do you write and interpret chemical equations for this class of reaction?

8. What can you say about the rates of precipitation reactions? How, for example, would you expect them to depend on temperature or concentration? How do they usually compare with other reaction rates?

9. What can you say about the macroscopic nature of the products of a precipitation (for example, the shape and size of crystals) and how it depends on reaction conditions?

10. What are some applications for this kind of reaction?

11.

12.

YOU WILL NEED TO KNOW

Concepts

1. The general principles of solubility, and how solubility may change with conditions (e.g., temperature) — Chapters 10, 11
2. The nature of electrolyte solutions (What is an electrotyte? How many ions form per mole of solute?) — Chapter 11
3. How to write and interpret [balanced] chemical equations — Chapter 3
4. How to define and interpret an equilibrium constant — Chapter 13

Math

1. How to perform stoichiometric calculations, particularly those involving concentration units — Chapters 3, 10 (and more worked examples in Chapter 16)
2. How to qualitatively and quantitatively predict the effects of changes in reactant or product concentrations on the position of equilibrium — Chapter 13
3. How to write the equilibrium constant expression for any reaction — Chapter 13

CHAPTER SUMMARY

This is the first of six chapters devoted to a discussion of the reactions of ions and molecules in water solution. Of the various types of reactions we will consider (precipitation, acid-base, complex ion formation, oxidation-reduction), precipitation is perhaps the simplest. A precipitation reaction will occur when two electrolyte solutions are mixed if one of the possible products is insoluble. To illustrate: when solutions of $MgCl_2$ and $NaOH$ are mixed, one of the two possible products, $Mg(OH)_2$, is insoluble (what is the other possible product?), so the following reaction occurs:

$$Mg^{2+}(aq) + 2OH^-(aq) \rightarrow Mg(OH)_2 (s)$$

Notice that the equation written for the reaction includes only those species which actually participate in it; "spectator" ions (Na^+, Cl^-) are omitted.

In order to decide when a precipitation reaction will occur, we need to know water solubilities of various electrolytes. Unfortunately, these cannot be predicted with any confidence from first principles. At the present state of electrolyte solution theory (Chapter 11), the best we can do is to rationalize after the fact; e.g., "explain" why $Mg(OH)_2$ should be much less soluble than NaCl. We have to resort to solubility rules (based simply on observation) such as those listed in Table 16.1 to predict the spontaneity of precipitation reactions.

Table 16.1 lists those compounds which can be expected to precipitate when 0.1 M solutions of the corresponding ions are mixed. If the solutions used are more dilute, and indeed for any quantitative calculations, we must know the solubility product of the electrolyte involved. For $Mg(OH)_2$ we find experimentally that

$$K_{sp} = [Mg^{2+}] \times [OH^-]^2 = 1 \times 10^{-11}$$

This is the equilibrium constant expression for the reaction

$$Mg(OH)_2 (s) \rightleftarrows Mg^{2+}(aq) + 2 OH^-(aq)$$

and can be manipulated in every way like the equilibrium constants of Chapter 13. In words, this equation tells us that $Mg(OH)_2$ will precipitate from any solution in which the product of the concentration of Mg^{2+}, times that of OH^- squared, exceeds 1×10^{-11}. Thus, $Mg(OH)_2$ is sufficiently insoluble to precipitate not only when 0.1 M solutions of Mg^{2+} and OH^- are mixed, but also with 0.01 M or even 0.001 M solutions. (Would $Mg(OH)_2$ form if the concentrations of both ions were 0.0001 m?) A knowledge of the solubility product also enables us to carry out many other practical calculations.

Precipitation reactions are used for a variety of analytical and preparative purposes. For example, we take advantage of the water insolubility of $Mg(OH)_2$ to:

a) detect Mg^{2+} in a mixture with other cations, none of which forms an insoluble hydroxide;

b) analyze quantitatively for Mg^{2+} by weighing the $Mg(OH)_2$ produced (or the MgO formed by heating it) when an excess of OH^- is added to a solution;

c) separate Mg^{2+} from sea water (this is the ultimate source of the magnesium metal in use today);

d) convert a soluble hydroxide (e.g., CsOH) to another soluble salt of that cation (e.g., CsCl, Cs_2SO_4) by adding the appropriate salt of Mg^{2+} (e.g., $MgCl_2$, $MgSO_4$), filtering off the precipitated $Mg(OH)_2$ and evaporating the solution remaining.

BASIC SKILLS

1. Write net ionic equations to represent precipitation reactions.

The reasoning involved here is described in some detail in Example 16.2 and applied in Problems 16.1 and 16.18. Note that in order to follow the example or work the problems you must be familiar with the solubility rules given in Table 16.1.

The phrase "net ionic equation" introduced in this chapter is perhaps a misnomer; "net equation" might be better. Such equations can involve the formulas of insoluble solids (e.g., $PbCl_2$) and, as we shall see in later chapters, molecules in water solution (e.g., HF, H_2O). The essential point is that in any equation of this type, *only those species which actually take part in the reaction are included in the equation.*

2. Given the equation for a precipitation reaction, relate the amounts (numbers of moles or grams) of reactants and products.

This skill is essentially identical with that first discussed in Chapter 3 in connection with mass relations in chemical reactions. Here for the first time it is applied to reactions in solution. The only difference is that, in solution reactions, you frequently need to convert concentrations to numbers of moles (or vice versa) using the definition of molarity introduced in Chapter 10:

$$molarity = \text{no. of moles solute/no. of liters solution}$$

Typical calculations of this type are shown in Example 16.1. Problems 16.5 and 16.22 are entirely analogous to this example. Problems 16.6 and 16.23 look somewhat different but follow the same principles. Indeed, they are considerably simpler than 16.5 and 16.22, provided you haven't forgotten how to convert grams to moles!

3. Given the formula of an ionic compound, write the expression for K_{sp}.

The principle is the same as that introduced in Chapter 13 in connection with K_c. Thus we have:

$$AgCl(s) \rightleftharpoons Ag^+(aq) + Cl^-(aq); \quad K_{sp} = [Ag^+] \times [Cl^-]$$

$$PbCl_2(s) \rightleftharpoons Pb^{2+}(aq) + 2Cl^-(aq); \quad K_{sp} = [Pb^{2+}] \times [Cl^-]^2$$

$$FeF_3(s) \rightleftharpoons Fe^{3+}(aq) + 3F^-(aq); \quad K_{sp} = [Fe^{3+}] \times [F^-]^3$$

$$As_2S_3(s) \rightleftharpoons 2As^{3+}(aq) + 3S^{2-}(aq); \quad K_{sp} = [As^{3+}]^2 \times [S^{2-}]^3$$

Note that here, as always, solids are omitted from the equilibrium constant expression.

4. Given the solubility in moles/liter of an ionic compound, calculate K_{sp}. Carry out the reverse calculation.

Both of these calculations are shown in Example 16.3. Note that the relationship between these two quantities depends upon the type of salt, as shown in the following table (where S = solubility).

Type of Salt	MX	MX_2 or M_2X	MX_3 or M_3X	M_2X_3 or M_3X_2
K_{sp}	S^2	$4S^3$	$27S^4$	$108S^5$
Example	AgCl	$PbCl_2$, Ag_2S	FeF_3	As_2S_3

The conversion from solubility to K_{sp} (and from grams to moles) is required in Problems 16.7 and 16.24. The reverse calculation is involved in Problems 16.8 and 16.25.

5. Use the value of K_{sp} to:

a) determine the concentration of an ion in solution, given that of the other ion in equilibrium with it.

— —

The solubility product of $PbCl_2$ is 1.7×10^{-5}. What is the concentration of Pb^{2+} in equilibrium with 0.020 M Cl^-? _____
Substituting directly into the expression for K_{sp}:

$$1.7 \times 10^{-5} = [Pb^{2+}] \times [Cl^-]^2 = [Pb^{2+}] \times (2.0 \times 10^{-2})^2$$

Solving:

$$[Pb^{2+}] = \frac{1.7 \times 10^{-5}}{(2.0 \times 10^{-2})^2} = \frac{1.7 \times 10^{-5}}{4.0 \times 10^{-4}} = 0.42 \times 10^{-1} = 0.042 \text{ M}$$

— —

This skill is often involved as one step in a more complex problem. Consider, for instance, Example 16.5, in which the ultimate aim is to obtain the percentage of Mg^{2+} left in sea water after adding OH^-; Problems 16.10, 16.11, and 16.27 are similar.

b) decide whether or not a precipitate will form when two solutions are mixed.

The principle here is a very simple one: a precipitate will form only if the concentration product exceeds K_{sp}. Otherwise, the solution will be unsaturated with respect to the solid, and solubility equilibrium will not be established. The skill is illustrated in Example 16.4 and Problems 16.9 and 16.26. You may notice a common assumption ordinarily made in problems of this type: when two solutions are mixed, it is assumed that there is no net change in volume (i.e., 400 ml of solution 1 mixed with 800 ml of solution 2 gives a final volume of 1200 ml).

c) decide which of two possible precipitates will form when two solutions are mixed.

This skill is involved in the first part of Problems 16.12 and 16.29. The general principle is that the precipitate which requires the lowest concentration of reagent will form first. Thus, in Problem 16.12 we calculate that the concentrations of Cl^- required to start to precipitate $AgCl$ and $PbCl_2$ are:

$$AgCl: \quad [Cl^-] \quad = K_{sp} \, AgCl/[Ag^+] \quad = 1.6 \times 10^{-10}/1.0 \times 10^{-2} = 1.6 \times 10^{-8}$$

$$PbCl_2: [Cl^-]^2 = K_{sp} \, PbCl_2 \, /[Pb^{2+}] = 1.7 \times 10^{-5}/1.0 \times 10^{-1} = 1.7 \times 10^{-4}$$

$$[Cl^-] \quad = (1.7 \times 10^{-4})^{1/2} \quad = 1.3 \times 10^{-2}$$

Since $1.6 \times 10^{-8} < 1.3 \times 10^{-2}$, we deduce that solid $AgCl$ forms first.

6. Calculate the percentage of a species present in a mixture, given analytical data for a precipitation reaction.

Example 16.6 illustrates a typical calculation of gravimetric analysis, in which a weighed sample is converted to a weighed precipitate. Here, all that is required is to establish the conversion factor which relates the weight of precipitate ($As_2 S_3$) to the weight of the substance being analyzed for (As). Once this factor (245.8 g $As_2 S_3 \simeq 149.8$ g As) is established, the problem unravels. Indeed, it becomes just another example of the weight-weight conversions first introduced in Chapter 3. See Problems 16.13 and 16.30.

Example 16.7 is typical of calculations required in volumetric analysis. Here one measures the volume of a reagent of known concentration required to react with a sample of known mass. The first step is to calculate the number of moles of reagent used ($AgNO_3$) and then relate that to the number of moles of the species being analyzed for (Cl^-). Then one calculates the number of grams of that species, and finally its percentage in the sample. Problems 16.14 and 16.31 are entirely analogous to Example 16.7.

*7. Use a table of solubilities to develop a scheme for separating a mixture of ions.

Refer to Example 16.8; note particularly the discussion at the top of p. 443. See also Problems 16.15 and 16.32.

8. Use precipitation reactions to "convert" one ionic compound to another.

This skill is described on p. 444. The basic idea is a simple one. To convert a compound MX (e.g., CsCl) to another compound MY (e.g., $CsNO_3$), we add a reagent ($AgNO_3$) containing the desired anion Y and a cation which forms a precipitate with the anion we want to get rid of, X. By filtering off this precipitate, we effectively substitute Y for X.

– –

What reagent should be added to a solution of Na_2SO_4 to give, after filtration, a solution of NaCl? _____
Clearly the anion of the reagent must be Cl^-. The cation should be one that gives a precipitate with SO_4^{2-}. A suitable choice would be Ba^{2+}, since $BaSO_4$ is insoluble. In other words, we would add a solution of $BaCl_2$ and filter off the $BaSO_4$ to obtain a solution of NaCl.

– –

See Problems 16.17 and 16.34. Note that in 16.17a, the desired product, NiS, is itself insoluble in water.

Problems

1. Write net ionic equations to represent any precipitation reactions that take place when 0.10 M solutions of the following compounds are mixed.

   a) $AgNO_3$ + KCl   b) KOH + $CuSO_4$
   c) NaCl + KNO_3 d) $Ba(OH)_2$ + $CuSO_4$

2. 200 ml of 0.30 M $Pb(NO_3)_2$ is mixed with 300 ml of 0.30 M HCl; a precipitate of $PbCl_2$ is obtained: $Pb^{2+}(aq) + 2Cl^-(aq) \rightarrow PbCl_2(s)$.

   a) How many moles of Pb^{2+}, NO_3^-, H^+ and Cl^- are present originally?
   b) Which is the limiting reagent, Pb^{2+} or Cl^-?
   c) How many moles of Pb^{2+} and Cl^- are consumed in the reaction?
   d) How many moles of Pb^{2+}, NO_3^-, H^+ and Cl^- remain after the reaction is over?

3. Write expressions for K_{sp} for:

a) $CuCO_3$ b) Ag_2CrO_4 c) $Al(OH)_3$ d) $Zn_3(PO_4)_2$

4. Complete the following table.

Compound	K_{sp}	Solubility (moles/liter)
$BaCrO_4$	2.0×10^{-10}	_____
$Mg(OH)_2$	1.0×10^{-11}	_____
CoS	_____	3×10^{-11}
PbF_2	_____	2×10^{-3}

5. Given that K_{sp} for $Al(OH)_3$ is 5×10^{-33},

a) calculate the concentration of Al^{3+} when $[OH^-]$ is 1×10^{-7}.
b) will a precipitate form when conc. $Al^{3+} = 1 \times 10^{-5}$ and conc. $OH^- = 2 \times 10^{-9}$?
c) if OH^- is slowly added to a solution 0.10M in both Al^{3+} and Mg^{2+}, what precipitate will form first? (K_{sp} $Mg(OH)_2 = 1 \times 10^{-11}$)

6. Determine the percentage of Ca in each of the following samples. Sample (a), weighing 0.600 g, gives 0.700 g of $CaCO_3$ on precipitation with CO_3^{2-}.
Sample (b), weighing 0.200 g, requires 15.0 ml of 0.120 M Na_2CO_3 to react exactly with it.

*7. Devise a scheme for separating the cations from a solution which may contain Na^+, Ba^{2+}, Ag^+, and Cu^{2+}.

*8. What reagent should be added to convert:

a) K_2SO_4 to KCl?
b) K_2SO_4 to KNO_3?
c) $Cu(NO_3)_2$ to $NaNO_3$?

– – – – – – – – – – – –

9. 400 ml of 0.10 M $Al(NO_3)_3$ is added to 100 ml of 0.60 M NaOH.

a) Write a net ionic equation for the precipitation reaction that occurs.

b) Complete the following table.

Ion	no. moles orig.	change	no. moles final	final conc. (mole/liter)
Al^{3+}	_____	_____	_____	_____
NO_3^-	_____	_____	_____	_____
Na^+	_____	_____	_____	_____
OH^-	_____	_____	_____	_____

10. The solubility product of $CaCO_3$ is 5×10^{-9}.

a) What is the solubility of $CaCO_3$ in moles/liter? grams/liter?
b) Will a precipitate form if 500 ml of 1×10^{-4} M $Ca(NO_3)_2$ is mixed with 500 ml of 2×10^{-5} M Na_2CO_3?
c) What is the concentration of CO_3^{2-} in equilibrium with $[Ca^{2+}] = 8 \times 10^{-3}$ M?
d) If enough Na_2CO_3 is added to a solution originally 5×10^{-3} M in Ca^{2+} to make $[CO_3^{2-}] = 1 \times 10^{-4}$, what percentage of the Ca^{2+} is left in solution?

11. Analysis of a sample containing Br^- shows that 16.6 ml of 0.241 M $AgNO_3$ is required to react with 0.530 g of sample.

a) What is the percentage of Br^- in the sample?
b) How many grams of AgBr are formed in the reaction?

SELF-TEST

True or False

1. $MgSO_4$ is more soluble than $BaSO_4$ because the charge ( )
density is higher for Mg^{2+} than for Ba^{2+}.

2. Using the "charge density rule," we would expect CaF_2 to be ( )
less soluble than $CaCl_2$.

3. The solubility of AgCl in 0.10 M NaCl is greater than it is in ( )
water.

4. The solubility of AgCl in 0.10 M $NaNO_3$ is greater than it is ( )
in water.

5. Any two salts that have the same K_{sp} value will have the ( )
same solubility.

6. When equal volumes of 0.1 M solutions of M^+ and X^- are ( )
mixed, a precipitate forms if K_{sp} of MX is less than 0.01.

7. A precipitate will form when enough Cl⁻ is added to a ( )
solution 0.10 M in Pb^{2+} to make conc Cl⁻ = 0.010 M (K_{sp} $PbCl_2$ =
1.7×10^{-5}).

8. A salt MNO_3 can be prepared by a precipitation reaction ( )
which involves adding $AgNO_3$ to a solution of MCl.

Multiple Choice

9. When 200 ml of 0.10 M $BaCl_2$ is added to 100 ml of 0.30 M ( )
$Na_2 SO_4$, the number of moles of $BaSO_4$ precipitated is
　　　a) 0.010　　　b) 0.020　　　c) 0.030　　　d) 0.20

10. 200 ml of 0.10 M $NiCl_2$ is added to 100 ml of 0.20 M NaOH. ( )
The number of moles of Ni^{2+} left in solution after precipitation is
　　　a) 0　　　b) 0.010　　　c) 0.030　　　d) 0.10

11. When solutions of $Pb(NO_3)_2$ and $Na_2 SO_4$ are mixed, the ( )
precipitate that forms is
　　　a) $PbSO_4$　b) $NaNO_3$　c) $PbSO_4$ and $NaNO_3$　d) none

12. The net ionic equation for the reaction that occurs when ( )
solutions of $AgNO_3$ and $Na_2 CO_3$ are mixed is
　　　a) $2AgNO_3 (aq) + Na_2 CO_3 (aq) \rightarrow Ag_2 CO_3 (s) + 2NaNO_3 (aq)$
　　　b) $2 Ag^+(aq) + CO_3^{2-}(aq) \rightarrow Ag_2 CO_3 (s)$
　　　c) $Na^+(aq) + NO_3^-(aq) \rightarrow NaNO_3 (s)$
　　　d) no reaction

13. The solubility product of $PbCO_3$ is 1×10^{-12}. In a solution ( )
in which $[CO_3^{2-}]$ = 0.2 M, the equilibrium concentration of Pb^{2+} is
　　　a) 1×10^{-12} M　　　　　　　　b) 5×10^{-12} M
　　　c) 2×10^{-11} M　　　　　　　　d) 1×10^{-6} M

14. The solubility product expression for $As_2 S_3$ is: K_{sp} = ( )
　　　a) $[As^{3+}] \times [S^{2-}]$　　　　　　b) $[As^{3+}]^2 \times [S^{2-}]$
　　　c) $[As^{3+}]^3 \times [S^{2-}]^2$　　　　　d) none of these

15. In order to remove 90% of the Ag^+ from a solution originally ( )
0.10 M in Ag^+, the $[CrO_4^{2-}]$ (K_{sp} $Ag_2 CrO_4$ = 1×10^{-12}) must be
　　　a) 1.1×10^{-12}　b) 1×10^{-11}　c) 1×10^{-10}　d) 1×10^{-8}

16. For a salt of formula MX_2, the solubility, S, will be related to ( )
K_{sp} by
　　　a) $S = K_{sp}$　　b) $S^2 = K_{sp}$　　c) $2S^3 = K_{sp}$　　d) $4S^3 = K_{sp}$

17. CrO_4^{2-} is used as an indicator in the titration of Cl^- with Ag^+ ( )
because
  a) it is yellow, whereas Cl^- is colorless
  b) Ag_2CrO_4, unlike $AgCl$, is soluble in water
  c) Ag_2CrO_4 precipitates before $AgCl$
  d) Ag_2CrO_4 precipitates only when virtually all the Cl^- has reacted

18. To separate and identify the ions in the mixture: Pb^{2+}, Cu^{2+}, ( )
Mg^{2+}, one might add the reagents H_2S, HCl and $NaOH$. They should
be added in the order
  a) HCl, H_2S, $NaOH$ b) H_2S, HCl, $NaOH$
  c) HCl, $NaOH$, H_2S d) $NaOH$, H_2S, HCl

19. To prepare pure $RbCl$ from Rb_2SO_4, one might use a ( )
precipitation reaction in which the other reagent is
  a) $BaCl_2$ b) $BaSO_4$ c) $NaCl$ d) $RbCl$

20. Consider the anions Cl^-, SO_4^{2-}, CO_3^{2-}.

  a) Which form insoluble salts with Pb^{2+}?_____

  b) Which form insoluble salts with Ba^{2+}?_____

  c) Which decrease the water solubility of KOH? _____

  d) Which decrease the water solubility of $BaSO_4$? _____

SELF-TEST ANSWERS

  1. F   (Rather, anion - anion contact in $MgSO_4$.)
  2. T
  3. F   (Common ion effect.)
  4. T   (Ionic strength effect.)
  5. F   (Only if they have the same type formula, e.g., MX and AB.)
  6. F   (Dilution lowers conc to 0.05 M.)
  7. F   ($1 \times 10^{-5} < 1.7 \times 10^{-5}$)
  8. T
  9. b   (Ba^{2+} is limiting reactant.)
10. b
11. a
12. b   (Representing all that could be observed.)
13. b
14. d   ($[As^{3+}]^2 \times [S^{2-}]^3$.)
15. d   (Note that $[Ag^+] = 0.010$.)
16. d   ($[M] = S$ while $[X] = 2S$.)

17. d   (Indicators reveal the "completion" of reaction i.e., approaching 100%.)
18. a   (Any other order would precipitate two ions at once.)
19. a
20. a) all   b) $SO_4{}^{2-}$, $CO_3{}^{2-}$ c) none   d) $SO_4{}^{2-}$

SELECTED READINGS

Fullman, R. L., The Growth of Crystals, *Scientific American* (March 1955), pp. 74-80.
   Most precipitation reactions involve the formation of crystalline solids. How does a crystal grow? Read this.

For practice in working problems involving solubility equilibria, see the problem manuals listed in the Preface.

17

Acids and Bases

QUESTIONS TO GUIDE YOUR STUDY

1. What properties are characteristic of an acid? a base? Can you think of some common examples of each, other than "stomach acid"?

2. How are these properties explained in terms of atomic-molecular structure?

3. What happens when a substance dissolves in water (reacts with it?) to give an acidic solution? What happens when a solution forms which acts as a base? What species are present in such solutions?

4. Can you predict what species will give basic solutions; acidic solutions?

5. What is meant by the *strength* of an acid? How are differences in acid strength explained in terms of atomic-molecular structure? (Likewise for bases?)

6. How are such comparisons made quantitative?

7. How do you quantitatively describe a water solution of an acid or a base? (Recall that you have been able to specify the relative concentrations of species in gaseous systems at equilibrium and in saturated water solutions of slightly soluble salts.)

8. How would you experimentally arrive at such a description?

9. What role does water play in the formation of acidic and basic solutions? Is water necessary for the existence of an acid or base?

10. Why study acids and bases? Where do you encounter them?

11.

12.

YOU WILL NEED TO KNOW

Concepts

    1. How to write Lewis structures — Chapter 7
    2. How to predict molecular geometry and polarity — Chapter 7

Math

    1. How to find logs and antilogs — Appendix 4
    2. How to write the equilibrium constant expression for any reaction — Chapter 13
    3. How to perform stoichiometric calculations, particularly those involving concentration units — Chapters 3, 10 (and more worked examples in Chapter 16)
    4. How to qualitatively and quantitatively predict the effects of changes in reactant or product concentrations on the position of equilibrium — Chapter 13

CHAPTER SUMMARY

This chapter and Chapter 18 deal with the general topic of acid-base reactions. In this chapter we consider the properties of acids and bases; in Chapter 18 the reactions between them are discussed.

An acidic water solution is one in which there is an excess of H^+ ions; i.e., the concentration of H^+ is greater than that of OH^-. Since, for any water solution at 25°C, $[H^+] \times [OH^-] = 1.0 \times 10^{-14}$, it follows that in an acidic solution, $[H^+] > 10^{-7}$ M. By the same token, a basic water solution is one in which there are excess OH^- ions; i.e., $[OH^-] > [H^+]$ or $[OH^-] > 10^{-7}$ M. Acidity or basicity can also be expressed in terms of pH, defined by the equation pH = $-\log_{10} [H^+]$. A solution of pH 7 is said to be neutral; an acidic solution has a pH less than 7 while a basic solution has a pH greater than 7.

Any species which forms H^+ ions when added to water will form an acidic solution. From this point of view, the following would be classified as acids:

    1) A large number of molecules containing ionizable hydrogen atoms. These may be binary compounds (e.g., $HCl \rightarrow H^+ + Cl^-$) or oxyacids (e.g., $HNO_3 \rightarrow H^+ + NO_3^-$), in which the ionizable hydrogen atom is covalently bonded to oxygen.

    2) A few anions containing ionizable hydrogen atoms, of which the HSO_4^- ion is typical: $HSO_4^- \rightarrow H^+ + SO_4^{2-}$.

3) Many, indeed most, cations. A hydrated cation such as $Zn(H_2O)_4{}^{2+}$ can lose a proton to give an acidic solution $(Zn(H_2O)_4{}^{2+} \rightarrow Zn(H_2O)_3OH^+ + H^+)$. The only cations which show no tendency to react in this way are those of the 1A metals and the large ions of group 2A.

Species which form basic solutions when added to water include:

1) The hydroxides of the 1A and 2A metals $(NaOH(s) \rightarrow Na^+ + OH^-)$.
2) Ammonia $(NH_3 + H_2O \rightarrow NH_4{}^+ + OH^-)$ and its organic derivatives such as CH_3NH_2.
3) Many, indeed most, anions, of which the F^- ion is typical:
$$F^- + H_2O \rightarrow HF + OH^-$$

A relatively few *strong acids* ionize completely when added to water. These include HCl, HBr, HI, HNO_3, $HClO_4$, and H_2SO_4 (1st ionization). Species which ionize only partially, called *weak acids*, are much more common. The strength of an acid is directly related to the magnitude of its equilibrium constant for ionization, K_a:

$$HX(aq) \rightleftharpoons H^+(aq) + X^-(aq); \quad K_a = \frac{[H^+] \times [X^-]}{[HX]}$$

The only strong bases are the hydroxides of the 1A and 2A metals, which are virtually completely ionized in water. Ammonia and its organic derivatives are weak bases, in the sense that the reaction

$$NH_3(aq) + H_2O \rightleftharpoons NH_4{}^+(aq) + OH^-(aq); \quad K_b = \frac{[NH_4{}^+] \times [OH^-]}{[NH_3]}$$

is incomplete. Anions derived from weak acids (e.g., F^-, CN^-, $C_2H_3O_2{}^-$, $CO_3{}^{2-}$) act as weak bases in water solution. The dissociation constant, K_b, of any weak base can be calculated by applying the Multiple Equilibrium Rule to show that

$$K_b \times K_a = K_w = 1.0 \times 10^{-14}$$

where K_a is the dissociation constant of the corresponding acid, often called the conjugate weak acid.

The relative strengths of different oxyacids can be estimated from structural considerations. Among a series of acids derived from the same element (e.g., $HClO_4$, $HClO_3$, $HClO_2$, $HClO$), acid strength increases with the number of oxygen atoms attached to the central atom. Among acids of the same type formula derived from different nonmetals (e.g., H_2SO_3, H_2SeO_3, H_2TeO_3), acid strength increases with the electronegativity of the central atom. Increasing acid strength results from decreasing the H—O bond strength by the removal of electrons from that bond.

Throughout most of this chapter, we have, at least by implication, used the Arrhenius definition of acids and bases as species which upon addition to water give H^+ or OH^- ions, respectively. The Brönsted-Lowry definition is somewhat broader; here an acid is taken to be a proton donor while a base is a proton acceptor. Thus, for the reaction:

$$NH_3\,(aq) + H_2O \rightleftharpoons NH_4^+(aq) + OH^-(aq)$$

the species NH_3 and OH^- are acting as Brönsted-Lowry bases while H_2O and NH_4^+ are acids. A still more general definition due to G. N. Lewis takes an acid to be an electron-pair acceptor and a base an electron-pair donor. Metal cations such as Zn^{2+} act as Lewis acids when they accept pairs of electrons from H_2O or NH_3 molecules to form the complex ions $Zn(H_2O)_4^{2+}$ and $Zn(NH_3)_4^{2+}$.

BASIC SKILLS

Students often find the material in this chapter rather difficult to master, principally because a large number of different concepts are presented. Some of these are quantitative in nature; others are qualitative. These are grouped separately below.

Quantitative Skills

1. Given one of the three quantities for a water solution, $[H^+]$, $[OH^-]$, and pH, calculate the other two quantities.

The basic relationships are:

$$pH = -\log_{10}[H^+]\ ;\ [H^+] \times [OH^-] = 1.0 \times 10^{-14}$$

The use of these relations is shown in Example 17.1. Note in part (a) the two alternatives for calculating pH. The second method, which makes use of the fact that $-\log x = \log 1/x$, is sometimes preferred by students who are wary of negative logarithms.

This skill is used in Problems 17.2 and 17.22 ($[H^+] \rightarrow$ pH), Problems 17.3 and 17.23 (pH $\rightarrow [H^+] \rightarrow [OH^-]$), and Problems 17.4 and 17.24. In the last two problems, you must first determine $[H^+]$ or $[OH^-]$ from the information given. You can assume that all the acids and bases listed in these problems are "strong," i.e., completely dissociated in water.

2. Calculate K_a for a weak acid, given:

a) $[H^+]$ in a solution prepared by dissolving HA in water to give a known initial concentration of HA.

This calculation is illustrated in Example 17.3. Note the two basic relationships which must apply when the solution is formed simply by dissolving HA in water:

$$[H^+] = [A^-] \; ; \; [HA] = \text{orig. conc. HA} - [H^+]$$

Problem 17.9 is similar to Example 17.3, except that you first have to calculate $[H^+]$ from the pH given. Problem 17.29 is similar, with one additional complication; you must obtain the original concentration of weak acid, knowing the number of grams per liter.

b) $[H^+]$ in a solution containing a known ratio of $[A^-]$ to $[HA]$.

- -

What is K_a for a weak acid if $[H^+] = 2.5 \times 10^{-4}$ when $[A^-] = 2[HA]$?

Setting up the expression for K_a:

$$K_a = \frac{[H^+] \times [A^-]}{[HA]} = 2.5 \times 10^{-4} \times \frac{[A^-]}{[HA]}$$

But, since $[A^-] = 2[HA]$, we find $[A^-]/[HA] = 2$, and:

$$K_a = (2.5 \times 10^{-4}) \times 2 = 5.0 \times 10^{-4}$$

- -

Example 17.4 shows how the ratio $[A^-]/[HA]$ can be set experimentally to a fixed value, in this case 1. Problem 17.10 is similar.

3. Given the original concentration of a weak acid and the value of K_a, calculate $[H^+]$ or per cent dissociation.

The calculation of $[H^+]$ is discussed in considerable detail in Examples 17.5 and 17.6. In the first example, the concentration of H^+ is so small compared to the original concentration of weak acid that the approximation

$$[HA] = \text{orig. conc. HA} - [H^+] \approx \text{orig. conc. } [HA]$$

is entirely justified. In contrast, in Example 17.6 the $[H^+]$ calculated is an appreciable fraction (9.2%) of the original concentration of weak acid, too large to be ignored. For that reason, it is desirable to refine the calculation, using a second approximation more nearly valid than the first.

The per cent dissociation of a weak acid is readily calculated from the defining relation:

$$\% \text{ diss.} = \frac{[H^+]}{\text{orig. conc. HA}}$$

You should be able to demonstrate for yourself that the per cent dissociation is 0.42% in Example 17.5 and 8.7% in Example 17.6.

Problems 17.11 and 17.31 illustrate the use of this skill (17.31d should read $NH_4 NO_3$). The quantities pH and [OH⁻] are of course calculated from [H⁺] and K_w. Values of K_a for the weak acids involved (HI is a strong acid) can be found in Table 17.5.

4. Relate the value of K_a for a weak acid to K_b for its conjugate base.

The relation here is:

$$K_b \text{ of } A^- = (1.0 \times 10^{-14})/K_a \text{ of HA}$$

See Example 17.7 and Problems 17.13 and 17.33.

5. Given the original concentration of a weak base and the value of K_b, calculate [OH⁻] and the per cent dissociation.

The calculations here are entirely analogous to those for weak acids. Compare Example 17.9 to Example 17.5; Problem 17.12 to 17.11; and Problem 17.32 to 17.31. Note that in Problems 17.12 and 17.32, a couple of "ringers" (strong bases) are included. "Other solute species" may be either ions or molecules; in Problem 17.12c, for example, they would include K⁺, CN⁻, and HCN.

Qualitative Skills

6. Predict whether a given acid or base is strong or weak.

While this kind of prediction is not called for directly in any of the problems, it is involved indirectly in several (e.g., Problems 17.14 and 17.34). There are only six common strong acids: HCl, HBr, HI, HNO_3, $HClO_4$ and $H_2 SO_4$. All other acids that you encounter in this course can be assumed to be weak. The strong bases are the hydroxides of the 1A and 2A metals (e.g., NaOH, Ca(OH)₂); all other bases can be assumed to be weak.

7. Predict whether a particular ion will be neutral, acidic, or basic in water solution.

Predictions of this sort can be made with the aid of Table 17.6. This table may seem formidable at first glance, but the principle behind it is really quite simple. The neutral anions are those derived from strong acids; the neutral cations are those derived from strong bases. Anions derived from weak acids are basic; cations derived from weak bases are acidic. There are essentially infinite numbers of ions in these two categories; those listed in the table as "basic anions" and "acidic cations" are only typical examples. Finally, there are a couple of oddball anions (HSO_4^- and $H_2PO_4^-$) which are acidic because they contain an ionizable proton.

This skill is involved in Problems 17.7 and 17.27; before you can write a balanced net ionic equation, you have to decide whether the ion in question gives a neutral, acidic, or basic water solution.

8. **Predict whether a given ionic compound will give a neutral, basic, or acidic water solution.**

As Example 17.8 implies, there are four categories to be considered:

- neutral cation, neutral anion: neutral solution
- neutral cation, basic anion: basic solution
- acidic cation, neutral anion: acidic solution
- acidic cation, basic anion: may be acidic, basic, or neutral (see NH_4F in Example 17.8)

This skill is involved in Problems 17.8 and 17.28.

9. **Write a net ionic equation to explain why a species gives an acidic or a basic solution.**

Before attempting to write such an equation, you first have to decide whether the solution produced is acidic, basic, or neutral. If you decide it is neutral, there is no reaction to explain and no equation to write. If the solution is acidic, one product must be H^+. The other product is simply the conjugate base of the reactant. Examples include:

$$HF(aq) \rightarrow H^+(aq) + F^-(aq)$$

$$HSO_4^-(aq) \rightarrow H^+(aq) + SO_4^{2-}(aq)$$

$$NH_4^+(aq) \rightarrow H^+(aq) + NH_3(aq)$$

$$Zn(H_2O)_4^{2+}(aq) \rightarrow H^+(aq) + Zn(H_2O)_3(OH)^+(aq)$$

In order to make a solution basic, a species must ordinarily pick up a proton from a water molecule, leaving an OH⁻ ion behind. Thus, to explain the fact that solutions containing NH_3, F^-, and CO_3^{2-} are basic, we would write:

$$NH_3\,(aq) + H_2O \rightarrow NH_4^{+}(aq) + OH^-(aq)$$

$$F^-(aq) + H_2O \rightarrow HF(aq) + OH^-(aq)$$

$$CO_3^{2-}(aq) + H_2O \rightarrow HCO_3^-(aq) + OH^-(aq)$$

Following these examples, you should be able to write the equations called for in Problems 17.7, 17.8, 17.27, and 17.28.

10. Predict the relative strengths of different oxyacids.

There are two simple principles here. The strength of an oxyacid increases with:

a) the electronegativity of the central atom

$$(HClO > HBrO > HIO);$$

b) the number of oxygen (or other highly electronegative) atoms attached to the central atom $(HClO_4 > HClO_3 > HClO_2 > HClO)$.

These principles can be applied to those parts of Problems 17.6 and 17.26 that involve oxyacids. A note of caution; don't attempt to apply these rules to species other than oxyacids, such as NH_4^{+} or H_2S.

11. Given the equation for an acid-base reaction, select the Brönsted acid and Brönsted base; the Lewis acid and Lewis base.

This skill requires only that you apply the definitions of Brönsted or Lewis acids and bases given on pp. 469 and 472 of the text. See Problems 17.16 and 17.36. In Problems 17.18 and 17.38, you are required to identify potential Brönsted acids (species that are capable of giving up a proton), Brönsted bases (species that can accept a proton), Lewis acids (species that can accept an electron pair), and Lewis bases (species that can donate an electron pair). Here, it is helpful to start by writing the Lewis structure of the species. In the case of H_2O, for example, we see from its structure that it could lose a proton to form OH^-, gain a proton to form H_3O^+, or donate a pair of unshared electrons. It could not, however, accept a pair of electrons, because there is no place to put them.

One precaution: some species which appear in principle to be capable of acting as acids or bases never (or almost never) do so. It might seem, for example, that SO_2 could act as a Lewis base, yet in practice it shows little if any tendency to donate one of its unshared electron pairs.

Problems

1. Complete the following table.

$[H^+]$	$[OH^-]$	pH
1.0×10^{-4}	_____	_____
_____	2.0×10^{-6}	_____
_____	_____	3.30

2. Calculate K_a for each of the following acids.

   a) When 0.100 mole of HB is dissolved to give one liter of solution, $[H^+]$ is found to be 3.0×10^{-4}.
   b) The concentration of H^+ is 2.5×10^{-5} in a solution in which $[HA] = [A^-]$.

3. What is the concentration of H^+ and the per cent dissociation in

   a) 1.0 M HF ($K_a = 7.0 \times 10^{-4}$)
   b) 0.10 M HF (make a second approximation)

4. Calculate K_b for F^-, given that K_a for HF is 7.0×10^{-4}.

5. Using the value of K_b obtained in (4), calculate $[OH^-]$ in a solution 0.10 M in F^-.

6. Classify each of the following as a strong acid, strong base, weak acid, or weak base.

   a) $Ba(OH)_2$ b) HNO_3 c) HF   d) F^- e) NH_4^+

7. Which of the following ions will act as acids in water solution? as bases? Which will be neutral?

   a) Na^+ b) Al^{3+} c) Fe^{3+} d) NO_3^- e) CO_3^{2-} f) PO_4^{3-}

8. Based on your answers to (7), predict whether the following salts will give acidic, basic, or neutral water solutions.

   a) $NaNO_3$ b) $Al(NO_3)_3$ c) Na_2CO_3 d) Na_3PO_4 e) $Fe(NO_3)_3$

9. Write net ionic equations to explain why

   a) a water solution of HCl is acidic.
   b) a water solution of CO_3^{2-} is basic.
   c) a water solution of NH_4^+ is acidic.
   d) a water solution of NH_3 is basic.

10. Arrange the following acids in order of increasing strength: H_2SeO_4, H_2SeO_3, H_2SO_4.

11. a) Identify the Brönsted acids and bases in the following reactions.

   $$HCO_3^-(aq) + OH^-(aq) \rightarrow CO_3^{2-}(aq) + H_2O$$

   $$HCO_3^-(aq) + H_2O \rightarrow H_2CO_3(aq) + OH^-(aq)$$

   $$HCO_3^-(aq) + H_2O \rightarrow CO_3^{2-}(aq) + H_3O^+(aq)$$

   b) Identify the Lewis acids and bases in the following reactions.

   $$Ag^+(aq) + 2NH_3(aq) \rightarrow Ag(NH_3)_2^+(aq)$$

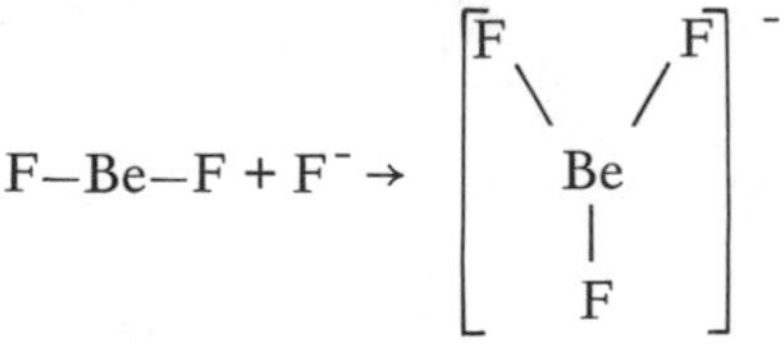

   $$F-Be-F + F^- \rightarrow \begin{bmatrix} F & & F \\ & \diagdown \;/ & \\ & Be & \\ & | & \\ & F & \end{bmatrix}^-$$

---

12. Consider the weak acid HCN ($K_a = 4.0 \times 10^{-10}$). Calculate:

   a) $[H^+]$ in a solution in which $[CN^-] = 0.010 \times [HCN]$
   b) $[H^+]$ in a solution prepared by dissolving 0.50 mole of HCN to give a liter of solution.
   c) the per cent dissociation in (b).
   d) the pH and $[OH^-]$ in (b).
   e) K_b for CN^-
   f) $[OH^-]$ in a solution 0.10 M in CN^-
   g) $[H^+]$ and the pH in (f)

13. Consider the following species:

$$KOH, HNO_2, HBr, NO_2^-, Br^-, CO_3^{2-}, K^+, Cu(H_2O)_4^{2+}$$

a) Which of these species are strong acids? strong bases? weak acids? weak bases? neutral?
b) Write balanced net ionic equations for the reactions in water of each of the weak acids.
c) Write balanced net ionic equations for the reaction with water of each of the weak bases.
d) For each of the reactions in (c), identify the Brönsted acids and bases.
e) Using the ions listed, give the formula of a salt that would give a neutral water solution; a basic water solution; an acidic water solution.
f) Give the formula of an oxyacid that would be stronger than HNO_2; weaker than HNO_2.

SELF-TEST

True or False

1. In a basic water solution at 25°C, $[H^+] > 10^{-7}$. ( )

2. It is impossible to have a solution with a negative pH. ( )

3. The pH of a solution prepared by dissolving 1×10^{-9} moles of HCl in a liter of water is 9. ( )

4. A solution which gives off bubbles when Na_2CO_3 is added is acidic. ( )

5. A solution which is colorless to phenolphthalein (Table 17.3) must be acidic. ( )

6. There are more weak acids than strong acids. ( )

7. Solutions containing the CN^- ion are expected to be basic. ( )

8. HF ($K_a = 7 \times 10^{-4}$) is a stronger acid than HNO_2 ($K_a = 4 \times 10^{-4}$). ( )

9. A solution of NH_4Cl is basic. ( )

10. The conjugate anions of strong acids are strong bases. ( )

Multiple Choice

11. The concentration of H^+ in a solution is 2×10^{-4} M. The OH^- ( )
is
    a) 2×10^{-4} M   b) 1×10^{-10} M
    c) 2×10^{-10} M   d) 5×10^{-11} M

12. The pH of the solution in Question 11 is ( )
    a) 3.0   b) 3.7   c) 4.0   d) 10.3

13. The solution referred to in Questions 11 and 12 ( )
    a) is acidic   b) is basic   c) is neutral   d) cannot exist

14. The pH of a solution is 5.5. The concentration of H^+ is about ( )
    a) 1×10^{-6} M   b) 1×10^{-5} M
    c) 3×10^{-5} M   d) 3×10^{-6} M

15. A 0.10 M solution of HCl would have a pH of ( )
    a) 0   b) 1.0   c) 7.0   d) 13.0

16. The pH of a 0.10 M solution of a weak acid would be ( )
    a) less than 1   b) 1   c) greater than 1   d) cannot say

17. Which one of the following is *not* a strong acid? ( )
    a) HCl   b) HF   c) HNO_3 d) $HClO_4$

18. Which one of the following is a strong base? ( )
    a) $Al(OH)_3$ b) NH_3 c) C_2H_5OH d) NaOH

19. Which one of the following ions upon addition to water ( )
would give a basic solution?
    a) NH_4^+ b) Na^+ c) $C_2H_3O_2^-$ d) NO_3^-

20. A certain weak acid is 10% dissociated in 1.0 M solution. In ( )
0.10 M solution, the percentage of dissociation would be
    a) greater than 10   b) 10   c) less than 10   d) complete

21. K_b for NH_3 is 2×10^{-5}. K_a for the NH_4^+ ion is ( )
    a) 2×10^{-5} b) 5×10^{-9}
    c) 5×10^{-10} d) 2×10^{-19}

22. Of the following acids, which is the strongest? ( )
    a) H_2TeO_4 b) H_2SeO_3 c) H_2SeO_4 d) H_2SO_4

23. In the reversible reaction: ( )
$HCO_3^-(aq) + OH^-(aq) \rightleftharpoons CO_3^{2-}(aq) + H_2O$, the Brönsted acids are
    a) HCO_3^- and CO_3^{2-} b) HCO_3^- and H_2O
    c) OH^- and H_2O d) OH^- and CO_3^{2-}

24. In the reaction: $BF_3 + NH_3 \rightarrow F_3B{:}NH_3$, BF_3 accepts an ( )
electron pair and acts as
    a) an Arrhenius base   b) a Brönsted acid
    c) a Lewis acid   d) a Lewis base

25. For the reaction: $HPO_4^{2-}(aq) + H_2O \rightarrow H_2PO_4^-(aq) + OH^-(aq)$ ( )
    a) $HPO_4{}^{2-}$ is an acid and OH^- its conjugate base
    b) H_2O is an acid and OH^- its conjugate base
    c) $HPO_4{}^{2-}$ is an acid and $H_2PO_4{}^-$ its conjugate base
    d) H_2O is an acid and $HPO_4{}^{2-}$ its conjugate base

26. Consider the species: Cu^{2+}, F^-, H_2O, $NH_4{}^+$:

    a) which, on addition to water, give acidic solutions? _____

    b) which, on addition to water, give basic solutions? _____

    c) which can act as Brönsted acids? _____

    d) which can act as Brönsted bases? _____

    e) which can act as Lewis acids? _____

    f) which can act as Lewis bases? _____

27. Which one of the following might you expect to be an ( )
"active ingredient" in Brand X Antacid?
    a) KOH   b) $SO_2(OH)_2$ c) $NaHCO_3$ d) NH_4Cl

SELF-TEST ANSWERS

    1. F
    2. F   (E.g., 10 M HCl.)
    3. F   (Where did the OH^- ions come from? See the Readings.)
    4. T   (Either that or it is supersaturated with gas!)
    5. F   (Could, for example, have a pH of 7.5.)
    6. T
    7. T   (Its conjugate acid, HCN, is weak.)
    8. T   (Caution: direct comparison of K_a's requires similar formulas;
        e.g., HA and HB.)
    9. F   ($NH_4{}^+$ is acidic.)
  10. F   (Cl^-, for example, is neutral.)
  11. d
  12. b
  13. a
  14. d
  15. b
  16. c   (Incomplete dissociation.)

17. b

18. d (OH⁻ not readily formed in the alcohol, C_2H_5OH. Why?)

19. c

20. a (% dissociation greater for more dilute solution.)

21. c

22. d (With the weakest O—H bond.)

23. b

24. c

25. b

26. a) Cu^{2+}, NH_4^+ b) F^- c) H_2O, NH_4^+ d) F^-, H_2O
    e) Cu^{2+} f) F^-, H_2O

27. c (Bicarbonate of soda. KOH, strong base, would be corrosive! $SO_2(OH)_2$ is a tricky but informative representation of sulfuric acid.)

SELECTED READINGS

Chilton, T. H., *Strong Water: Nitric Acid, Its Sources, Methods of Manufacture, and Uses,* Cambridge, Mass., MIT Press, 1968.
    Readable discussion of the chemistry and the economics of production . . . for a very important industrial commodity (most of which is used in the preparation of fertilizers).
Jensen, W.B., Lewis Acid-Base Theory, *Chemistry* (March 1974), pp. 11-14.
    The first of a very readable series of articles on acid-base theory.
Mogul, P. H. and J. S. Schmuckler, Dilute Solutions of Strong Acids: The Effect of Water on pH, *Chemistry* (October 1969), pp. 14-17.
    What would be the pH of 1 × 10⁻⁹ M HCl (Question #3, Self-Test)? Water makes a contribution which must be counted when dealing with very dilute solutions of acids and bases.

Further practice in working problems can be gotten from the problem manuals listed in the Preface.

18

Acid-Base Reactions

QUESTIONS TO GUIDE YOUR STUDY

1. What reactions involve the participation of an acid or a base, or both? What, for example, constitutes a *neutralization*?

2. What effect does acid or base strength have on the nature of these reactions?

3. What energy effects are associated with acid-base reactions? Can you explain them in terms of what the molecules and ions are doing?

4. How can you predict the spontaneity of an acid-base reaction? its extent?

5. How would you experimentally follow an acid-base reaction? What would you measure to determine, for example, its extent?

6. What can be said about the rates of acid-base reactions; about reactions in which an acid or a base plays the part of a catalyst?

7. What are the properties of a buffer system? How do you account for them?

8. Where do you find buffers? (Can you name one?) What are some applications?

9. What are some of the applications of acid-base reactions?

10. How does an acid-base indicator work?

11.

12.

YOU WILL NEED TO KNOW

Concepts

1. How to predict whether a given species will act as an acid or a base; how to predict relative strengths of acids and bases; and, in general, most of the concepts introduced in the preceding chapter — see Chapter 17 and this section of the study guide, Chapter 17
2. How to write net ionic equations — Chapter 16

Math

1. How to work problems involving equilibrium constants in general — See Chapter 16 for solubility product constants; Chapter 17, for dissociation and hydrolysis constants
2. For other essential math, see this section of Chapter 17.

CHAPTER SUMMARY

The neutralization reaction:

$$H^+(aq) + OH^-(aq) \rightarrow H_2O; K = 10^{14}$$

takes place when any strong acid reacts with any strong base in water solution. If the acid is weak, it is more appropriate to write the equation as

$$HX(aq) + OH^-(aq) \rightarrow H_2O + X^-(aq)$$

since the principal species in solution before neutralization is the HX molecule rather than the H^+ ion. Equilibrium constants for the reactions of weak acids with strong bases are smaller than that for neutralization ($K = K_a \times 10^{14}$) but are ordinarily large enough to drive the reaction virtually to completion. A similar situation exists in the reaction of a weak base with a strong acid where $K = K_b \times 10^{14}$. Here, the weak base may be a molecule or ion in water solution (e.g., NH_3, CO_3^{2-}) or an anion of a water-insoluble salt:

$$BaCO_3(s) + 2 H^+(aq) \rightarrow Ba^{2+}(aq) + CO_2(g) + H_2O$$

Reactions of this type are commonly used to bring into solution salts containing the anions of weak acids (F^-, CO_3^{2-}, S^{2-}).

The concentration of an acid or base in water solution or in a solid mixture can be determined by titration, in which we measure the volume of a reagent of known concentration required to reach the equivalence point. The indicator used is ordinarily a weak acid. Ideally, we choose an indicator whose K_a is equal to the concentration of H^+ at the equivalence point. In a strong acid-strong base titration, the choice of indicator is not critical since $[H^+]$ changes rapidly near the equivalence point. For a weak acid or weak base, the pH changes much more slowly. Indeed, if an acid or base is very weak, the change in pH is so gradual that an ordinary acid-base titration cannot be carried out.

A solution containing a mixture of a weak acid and its conjugate weak base acts as a buffer in the sense that it prevents drastic change in pH when small amounts of strong acid or strong base are added. For a sodium acetate-acetic acid buffer, the reactions are:

addition of strong acid: $H^+(aq) + C_2H_3O_2^-(aq) \rightarrow HC_2H_3O_2(aq)$
addition of strong base: $OH^-(aq) + HC_2H_3O_2(aq) \rightarrow C_2H_3O_2^-(aq) + H_2O$

Ideally, we use a buffer system when the K_a of the weak acid is equal to the concentration of H^+ that we wish to maintain. Thus, we might use an $C_2H_3O_2^- - HC_2H_3O_2$ buffer ($K_a = 1.8 \times 10^{-5}$) to maintain pH in the range 4.5 to 5.0.

Acid-base reactions, like precipitation reactions discussed in Chapter 16, are widely used in chemical analysis and synthesis, often in commercial applications. Examples include:

a) Acid-base titrations to determine the per cent of HCO_3^- in a mixture.
b) Separation of a mixture of $BaCO_3$ and $BaSO_4$ by adding a strong acid to bring CO_3^{2-} into solution.
c) Preparation of volatile species such as CO_2 (add H^+ to $CaCO_3$) or NH_3 (add OH^- to a salt containing the NH_4^+ cation).
d) The Solvay process, by which $NaHCO_3$ and Na_2CO_3 are produced, ultimately from $CaCO_3$, $NaCl$ and H_2O.

BASIC SKILLS

This chapter, like Chapter 17, covers a rather large number of skills. Here, also, it is helpful to classify these skills into two categories.

Quantitative Skills

1. Use the Law of Multiple Equilibria to calculate K for an acid-base reaction.

This law, stated on p. 461, is particularly useful in this chapter. In order to use it, you must express the reaction in which you are interested as the sum of two or more reactions for which the equilibrium constants are known. This process is illustrated on p. 478, where the law is used to obtain a general expression for K for the reaction of a weak acid with OH^- ions; in Example 18.2 a numerical value of K is obtained for a particular weak acid,

$HC_2H_3O_2$. A more complicated case is considered on pp. 481-2, where, by adding three reactions, it is shown that for

$$BaCO_3 \text{ (s)} + 2H^+(aq) \rightarrow Ba^{2+}(aq) + H_2CO_3 \text{ (aq)}; \ K = \frac{K_{sp} BaCO_3}{K_1 K_2}$$

where K_1 and K_2 are respectively the first and second ionization constants of H_2CO_3. An entirely analogous expression can be derived for the case of an insoluble sulfide dissolving in strong acid; see the discussion at the top of p. 497 and Example 18.8 (K_1 and K_2 here refer to the first and second ionization constants of H_2S).

This skill is required to work Problems 18.3a, 18.4, 18.19a, 18.22a, and 18.23. The equilibrium constants derived using the law can, of course, be used for any of the various purposes discussed in Chapters 16 and 17. See, for instance, part b of Example 18.2; Problems 18.3b and 18.22b are analogous. Again, Example 18.3 shows how K for the reaction of $BaCO_3$ with strong acid can be used to calculate the solubility of this compound in acid solution. In Example 18.8, a quite similar application is made for the reactions of CoS and CuS with H^+. See also Problems 18.7, 18.19b, 18.26, and 18.38. The first step in each of these problems is to calculate the appropriate equilibrium constant. That constant is then used in latter parts of the problem.

2. Relate titration data for an acid-base reaction to:

a) the concentration of an acid or base in solution.

The calculations involved here are illustrated in Example 18.4. Problems 18.5 and 18.24 are similar, but note that in a titration with base, one mole of H_2SO_4 furnishes two moles of H^+.

b) the gram equivalent weight of an acid (i.e., the weight that reacts with one mole of OH⁻).

- -

A sample of a solid acid weighing 0.150 g requires 24.0 ml of a solution 0.200 M in OH⁻ for complete reaction. What is the GEW of the acid? _____
We start by calculating the number of moles of OH⁻ used.

no. moles OH⁻ = (0.0240 lit) (0.200 mole/lit) = 4.80×10^{-3} mole OH⁻

Since 0.150 g of acid reacts with 4.80×10^{-3} mole OH⁻, the weight that reacts with one mole of OH⁻ must be:

$$\frac{1.50 \times 10^{-1} \text{g acid}}{4.80 \times 10^{-3} \text{ mole OH}^-} = 0.312 \times 10^2 \frac{\text{g acid}}{\text{mole OH}^-} = 31.2 \text{ g acid/mole OH}^-$$

Hence, the GEW of the acid must be 31.2 g.

- -

Problem 18.8 is entirely analogous to the example above. In Problem 18.27, the calculation is reversed; knowing the GEW of KHPh and the volume of NaOH required to react with a weighed sample of the solid, you are asked to calculate the molarity of the NaOH.

3. **Given the composition of a buffer, determine its pH before and after the addition of known amounts of a strong acid or base.**

Any calculation of this sort proceeds through three steps:

a) the ratio $[HA]/[A^-]$ is first determined. For the original buffer, the amounts of HA and A^- are either given or readily calculated from the statement of the problem.

Addition of x moles of a strong acid to the buffer increases the number of moles of HA by x and decreases the number of moles of A^- by the same amount. Conversely, if x moles of a strong base are added, the number of moles of A^- increases by x while that of HA decreases by x. One point to keep in mind: since HA and A^- are present in the same solution, their concentrations must be in the same ratio as the numbers of moles, i.e.,

$$\frac{[HA]}{[A^-]} = \frac{\text{no. moles HA}}{\text{no. moles A}^-}$$

b) $[H^+]$ is calculated from the equation for K_a, i.e.,

$$[H^+] = K_a \times \frac{[HA]}{[A^-]}$$

c) pH is obtained from the defining equation: $pH = -\log_{10} [H^+]$

Typical calculations are shown in Example 18.7. Problems 18.13 and 18.32 are entirely analogous, except that in 18.13 you must realize that the addition of 0.20 mole of NaOH to 0.40 mole of $HCHO_2$ will give a solution containing 0.20 mole of $HCHO_2$ and 0.20 mole of CHO_2^-. Problem 18.14 is a little easier than 18.13, while 18.33 is somewhat more difficult.

4. Given the equation for an acid-base reaction, relate the normality of a reagent to its molarity, or its gram equivalent weight to its gram molecular weight.

This topic is discussed on p. 485. The general relations are:

$$GEW = GFW/n; \quad N = n \times M$$

where, for an acid: n = no. moles OH^- reacting with one mole of acid
for a base: n = no. moles H^+ reacting with one mole of base
See Problems 18.10 and 18.29.

Qualitative Skills

5. Write net ionic equations to describe acid-base reactions.

The nature of the equation depends upon whether the acid and base are strong or weak, and whether both species are in water solution or one is a solid.

a) *Strong acid – Strong base.* If both species are in solution, the equation is simply:

$$H^+(aq) + OH^-(aq) \rightarrow H_2O$$

If the reactant is a solid base, the equation is written to reflect that fact. Thus, for the addition of a strong acid to solid magnesium hydroxide we would write

$$2H^+(aq) + Mg(OH)_2(s) \rightarrow 2H_2O + Mg^{2+}(aq)$$

b) *Weak acid – Strong base.* For a reaction taking place in water solution, the reactants are the weak acid molecule or ion and the OH^- ion. The products are an H_2O molecule and the conjugate base of the weak acid. Typical examples include:

$$HC_2H_3O_2(aq) + OH^-(aq) \rightarrow H_2O + C_2H_3O_2^-(aq)$$

$$HCO_3^-(aq) + OH^-(aq) \rightarrow H_2O + CO_3^{2-}(aq)$$

$$NH_4^+(aq) + OH^-(aq) \rightarrow H_2O + NH_3(aq)$$

These are the equations that we would write to describe the reactions of a solution of NaOH with solutions of $HC_2H_3O_2$, $NaHCO_3$, and NH_4Cl respectively.

c) *Strong acid – Weak base.* In water solution the reactants are H^+ and the weak base (molecule or anion). The product is the conjugate acid of the weak base. For the reaction of a solution of any strong acid with solutions containing the NH_3 molecule, the S^{2-} ion or the CO_3^{2-} ion, we would write:

$$NH_3(aq) + H^+(aq) \rightarrow NH_4^+(aq)$$

$$S^{2-}(aq) + 2H^+(aq) \rightarrow H_2S(aq)$$

$$CO_3^{2-}(aq) + 2H^+(aq) \rightarrow H_2CO_3(aq) \rightleftharpoons CO_2(aq) + H_2O$$

Frequently in reactions of this type the "weak base" is an anion present as an insoluble solid (e.g., sulfide or carbonate). To represent the reaction of any strong acid with insoluble ZnS and $CaCO_3$ we would write:

$$ZnS(s) + 2H^+(aq) \rightarrow H_2S(aq) + Zn^{2+}(aq)$$

$$CaCO_3(s) + 2H^+(aq) \rightarrow H_2CO_3(aq) + Ca^{2+}(aq)$$
$$CO_2(aq) + H_2O$$

- -

Write balanced net ionic equations for the acid-base reactions that occur when

— a solution of HCl is added to $Ca(OH)_2(s)$. ———
— a solution of HNO_2 is added to a solution of NaOH ———
— a solution of HCl is added to $BaCO_3(s)$ ———

To write these equations, you must realize that HCl is a strong acid, entirely dissociated to H^+ and Cl^- in solution, while HNO_2 is a weak acid, present in solution primarily as the HNO_2 molecule.

$$2H^+(aq) + Ca(OH)_2(s) \rightarrow 2H_2O + Ca^{2+}(aq)$$

$$HNO_2(aq) + OH^-(aq) \rightarrow H_2O + NO_2^-(aq)$$

$$2H^+(aq) + BaCO_3(s) \rightarrow H_2O + CO_2(g) + Ba^{2+}(aq)$$

- -

In Problems 18.2 and 18.21, you have a chance to apply these principles in the simplest case, where all the species are in water solution.

6. Given the K_a values of several indicators, choose one which is appropriate for a particular acid-base reaction.

A general discussion of the principles involved is given on pp. 485-90. Summarizing this discussion:

— if both acid and base are strong, almost any indicator will work.
— if the acid is strong and the base is weak, use an indicator with a $K_a > 10^{-7}$ (e.g., methyl red).
— if the acid is weak and the base is strong, use an indicator with a $K_a < 10^{-7}$ (e.g., phenolphthalein).
— if both acid and base are weak, forget it; no indicator will work.

See Problems 18.9 and 18.28.

7. Use acid-base reactions to separate or distinguish between ions.

This skill is required in Problems 18.16, 18.17 and 18.35. (Note that in 18.35, both precipitation and acid-base reactions are allowed; in 18.17, anything goes!) Two helpful hints here:

—   Many acid-base reactions yield gases which are easily identified (CO_2, H_2S, SO_2, NH_3).
—   Insoluble solids containing a basic anion (e.g., CO_3^{2-}, S^{2-}, OH^-) can ordinarily be brought into solution by treating with a strong acid. If the anion is neutral (e.g., SO_4^{2-}, Cl^-), addition of a strong acid will have no effect.

8. Use acid-base reactions to prepare inorganic compounds.

This skill is required in Problems 18.15 and 18.34 (precipitation reactions are also allowed). If you are baffled by one of the parts of these problems, refer back to pp. 493-5, where the principles are discussed. A couple of hints:

—   Many of the reactions which are used to separate or identify ions also serve to convert one compound to another. Thus, treatment of $FeS(s)$ with hydrochloric acid gives off H_2S and leaves $FeCl_2$ in solution; after evaporation, you have, in effect, converted FeS to $FeCl_2$.
—   Simple neutralization is often a key step in a conversion. Obviously, NaOH could be converted to NaI by neutralization with HI followed by evaporation. In a less obvious case, you might convert $MgCl_2$ to MgI_2 by first precipitating $Mg(OH)_2$, then neutralizing the precipitate with HI, and finally evaporating the resulting solution.

Problems

1. Use the Law of Multiple Equilibria to calculate K for:

a) $HNO_2(aq) + OH^-(aq) \rightarrow NO_2^-(aq) + H_2O$; (add 2 equations)
b) $HF(aq) + NO_2^-(aq) \rightarrow F^-(aq) + HNO_2(aq)$; (add 2 equations)
c) $CaCO_3(s) + 2H^+(aq) \rightarrow Ca^{2+}(aq) + H_2CO_3(aq)$; (add 3 equations)

2. a) If 16.0 ml of 0.150 M NaOH is required to react with 12.0 ml of HCl, what is the molarity of the HCl?
   b) If 16.0 ml of 0.150 M NaOH is required to react with a pure sample of an acid weighing 0.500 g, what is the GEW of the acid?

3. A certain buffer contains 0.50 mole of a weak acid HA ($K_a = 1.0 \times 10^{-6}$) and 0.20 mole of its conjugate base, A^-.

a) What is the value of the ratio $[HA]/[A^-]$ in the buffer?
b) What is $[H^+]$ in the buffer?
c) What is the pH of the buffer?
d) What are the answers to a), b) and c) if 0.05 mole of OH^- is added to the buffer?

*4. Complete the following table.

	GEW H_2CO_3	M H_2CO_3	N H_2CO_3
$H_2CO_3(aq) + OH^-(aq) \rightarrow HCO_3^-(aq) + H_2O$	_____	1.0	_____
$H_2CO_3(aq) + 2OH^-(aq) \rightarrow CO_3^{2-}(aq)\ 2H_2O$	_____	_____	1.6

5. Write net ionic equations for the reaction between:

a) solutions of HNO_3 and KOH; a solution of HNO_3 and $Sr(OH)_2$ (s).
b) solutions of HF and KOH.
c) solutions of HNO_3 and NH_3; a solution of HNO_3 and $CaCO_3$ (s).

6. Given indicators with $K_a = 1 \times 10^{-5}$, 1×10^{-7}, and 1×10^{-9}, which one would you choose to titrate:

a) NH_3 with HCl?   b) HF with NaOH?
c) HCl with NaOH?   d) HF with NH_3?

7. Using acid-base reactions, distinguish between:

a) CO_3^{2-} and Br^-
b) NH_4^+ and K^+
c) HS^- and Cl^-

8. How would you carry out the following conversions, using acid-base reactions?

a) $Ca(NO_3)_2$ from $Ca(OH)_2$
b) $Mg(NO_3)_2$ from $MgCl_2$

9. Consider the reaction that occurs when excess strong acid is added to a solution containing $C_2H_3O_2^-$ ions.

 a) Write a balanced net ionic equation for this reaction.
 b) Calculate K for this reaction.
 c) Would phenolphthalein ($K_a = 10^{-9}$) be a suitable indicator for a titration of this sort?
 d) How many milliliters of 0.100 M H^+ must be added to 20.0 ml of 0.200 M $C_2H_3O_2^-$ to reach the equivalence point?
 e) What is the gram equivalent weight of $C_2H_3O_2^-$ in this reaction?

10. A buffer is to be prepared from a weak acid, HA ($K_a = 2.0 \times 10^{-4}$) and its salt, NaA. It is desired that the buffer have a pH of 4.00.

 a) What must [H^+] be in this buffer?
 b) What must be the value of the ratio [HA]/[A^-] in this buffer?
 c) How many moles of HA must be added to 1.0 mole of A^- to form this buffer?
 d) How many moles of H^+ must be added to 1.0 mole of A^- to form this buffer (note that H^+ converts A^- to HA)?

11. Consider the reaction that occurs when a strong acid is added to a water-insoluble sulfide, MS.

 a) Write a balanced net ionic equation for this reaction.
 b) Show that $K = K_{sp}MS/(K_1 H_2 S \times K_2 H_2 S)$
 c) Explain how this reaction could be used to distinguish MS from MSO_4.
 d) Explain how this reaction could be used to convert MS to MCl_2; to $M(NO_3)_2$.

SELF-TEST

True or False

1. The equilibrium constant for the reaction of a weak acid with ( )
a strong base is smaller than that for a strong acid-strong base reaction.

2. When 100 ml of 0.10 M $HC_2H_3O_2$ is added to 100 ml of ( )
0.10 M NaOH, a neutral solution is produced.

3. When 100 ml of 0.10 M HCl is added to 100 ml of 0.10 M ( )
NaOH, a neutral solution of NaCl is formed.

4. The equilibrium constant for the reaction of $C_2H_3O_2^-$ with   ( )
HCl is equal to $1/K_a$, where K_a is the dissociation constant for
$HC_2H_3O_2$.

5. The gram equivalent weight of acetic acid, $HC_2H_3O_2$, in an   ( )
acid-base titration, is always equal to its gram molecular weight.

6. At the equivalence point of the titration of a strong acid with   ( )
a weak base, the pH is greater than 7.

7. A mixture of 100 ml of 1.0 M HCl with 100 ml of 2.0 M   ( )
$NaC_2H_3O_2$ would act as a buffer.

8. A mixture of 100 ml of 1.0 M HCl with 100 ml of 1.0 M   ( )
$NaC_2H_3O_2$ would serve as a buffer.

9. Ammonia can be prepared by heating a solution of   ( )
ammonium chloride with a strong acid.

10. Sodium hydrogen carbonate could be prepared by saturating   ( )
a solution of NaOH with CO_2 and evaporating.

Multiple Choice

11. The equation for the reaction of a water solution of the weak   ( )
acid HF with a solution of NaOH is best written as
   a) $H^+(aq) + F^-(aq) \rightarrow HF(aq)$
   b) $H^+(aq) + OH^-(aq) \rightarrow H_2O$
   c) $HF(aq) + OH^-(aq) \rightarrow H_2O + F^-(aq)$
   d) $HF(aq) + NaOH(aq) \rightarrow H_2O + NaF(aq)$

12. The equation for the reaction of a water solution of ammonia   ( )
with a water solution of HCl is best written as
   a) $NH_3(aq) + H_2O \rightarrow NH_4^+(aq) + OH^-(aq)$
   b) $NH_4^+(aq) \rightarrow NH_3(aq) + H^+(aq)$
   c) $NH_3(aq) + H^+(aq) \rightarrow NH_4^+(aq)$
   d) $NH_3(aq) + HCl(aq) \rightarrow NH_4^+(aq) + Cl^-(aq)$

13. The equation for the reaction of Ag_2CO_3, a water insoluble   ( )
solid, with a strong acid is best written as
   a) $Ag_2CO_3(s) \rightarrow 2\ Ag^+(aq) + CO_3^{2-}(aq)$
   b) $CO_3^{2-}(aq) + 2\ H^+ \rightarrow CO_2(g) + H_2O$
   c) $Ag_2CO_3(s) \rightarrow Ag_2O(s) + CO_2(g)$
   d) $Ag_2CO_3(s) + 2\ H^+ \rightarrow 2\ Ag^+(aq) + CO_2(g) + H_2O$

14. A certain weak acid has a dissociation constant of 1×10^{-4}.   ( )
The equilibrium constant for its reaction with a strong base is:
   a) 1×10^{-4} b) 1×10^{-10} c) 1×10^{10} d) 1×10^{14}

15. When 500 ml of 0.10 M NaOH is reacted with 500 ml of 0.20   ( )
M HCl, the final concentration of H^+ is:

    a) 0.10 M   b) 0.20 M   c) 0.050 M   d) 10^{-7} M

16. Which one of the following salts will not be soluble in strong   ( )
acid?

    a) $PbCO_3$ b) CaF_2 c) $PbSO_4$ d) KCl

17. For the reaction: $H_2PO_4^-(aq) + 2\ OH^-(aq) \rightarrow PO_4^{3-}(aq)$ ( )
$+ 2\ H_2O$, the gram equivalent weight of $H_2PO_4^-$ is equal to

    a) 2 X GFW   b) GFW
    c) 1/2 GFW   d) that of PO_4^{3-}

18. When Na_2CO_3 takes part in an acid-base reaction, the ratio   ( )
of its normality to its molarity is

    a) 1/2   b) 1   c) either 1/2 or 1   d) either 1 or 2

19. A certain buffer contains equal concentrations of X^- and HX.   ( )
The K_b of X^- is 10^{-10}. The pH of the buffer is

    a) 4   b) 7   c) 10   d) 14

20. A buffer is formed by adding 500 ml of 0.20 M $HC_2H_3O_2$ to   ( )
500 ml of 0.10 M $NaC_2H_3O_2$. What is the maximum amount of HCl
that can be added to this solution without exceeding the capacity of
the buffer?

    a) 0.01 mole b) 0.05 mole c) 0.10 mole d) 0.20 mole

21. If one were to prepare a buffer by using HSO_3^- and SO_3^{2-},   ( )
its pH would probably be in the range of _____. (Ka $HSO_3^- = 3$ X
10^{-8})

    a) 6.5-8.5   b) 10-12   c) 2.0-4.0   d) 3 ± 1

22. $CuCl_2$ is prepared from $Cu(OH)_2$ by adding 0.10 M HCl to   ( )
0.20 mole of solid $Cu(OH)_2$. How much HCl should be added?

    a) 500 ml   b) 1000 ml
    c) 2000 ml   d) some other volume

23. The most appropriate equation for the reaction referred to in   ( )
Question 22 is:

    a) $Cu(OH)_2(s) + 2\ H^+(aq) \rightarrow Cu^{2+}(aq) + 2\ H_2O$
    b) $Cu^{2+}(aq) + 2\ Cl^-(aq) \rightarrow CuCl_2(s)$
    c) $H^+(aq) + OH^-(aq) \rightarrow H_2O$
    d) $Cu(H_2O)_4^{2+}(aq) \rightarrow Cu(H_2O)_3(OH)^+(aq) + H^+(aq)$

24. The metal sulfides CoS, CuS and FeS have solubility products   ( )
of 10^{-21}, 10^{-25} and 10^{-17}, respectively. The most soluble in acid will
be

    a) CoS   b) CuS   c) FeS   d) cannot say

25. Consider the three indicators: methyl red ($K_a = 10^{-5}$), litmus ($K_a = 10^{-7}$) and phenolphthalein ($K_a = 10^{-9}$). Which indicator would you use to titrate

    a) NH_3 with HCl? _____

    b) NaOH with HCl? _____

    c) $C_2H_3O_2^-$ with HNO_3? _____

    d) $HC_2H_2O_2$ with NaOH? _____

    e) HCN with NH_3? _____

SELF-TEST ANSWERS

    1. T

    2. F   (Solution contains $C_2H_3O_2^-$, hence is basic.)

    3. T

    4. T   (Net reaction is reverse of dissociation of $HC_2H_3O_2$.)

    5. T

    6. F   (Less than 7; solution will contain the weak conjugate acid.)

    7. T   (Equivalent amounts of $HC_2H_3O_2$, $C_2H_3O_2^-$ are present.)

    8. F   (No capacity to absorb H^+.)

    9. F   (Strong base.)

  10. T   (CO_2 gives acidic water solution.)

  11. c

  12. c

  13. d

  14. c

  15. c   (Don't forget the total volume.)

  16. c

  17. c

  18. d   (Depending on whether reacted to HCO_3^- or CO_2.)

  19. a

  20. b   (Equal to number of moles of weak base.)

  21. a

  22. d   (4000 ml; note that one mole of solid gives two moles of OH^-.)

  23. a   (Followed perhaps by evaporation – answer b.)

  24. c

  25. a) MR b) any one c) MR d) P e) none would work

SELECTED READINGS

See the readings list for Chapter 17.

For practice in acid-base stoichiometry, see the problem manuals listed in the Preface.

19

Complex Ions; Coordination Compounds

QUESTIONS TO GUIDE YOUR STUDY

1. What is a *complex ion;* a *coordination compound*? What special properties do they possess?

2. Where are you likely to encounter complex ions? What elements are most likely to form complexes?

3. What kind of experimental support do we have for the existence of complexes? How could we show that they exist in the solid state; in solution?

4. What is the nature of the bonding in these species? What geometries do you associate with various complexes?

5. How does the bonding and geometry of complexes account for their properties?

6. How do you account for the formation of a complex ion in terms of ΔH and ΔS?

7. How can you decide on the relative stabilities of complexes? How can you measure the extent to which one species is formed at the expense of another? How is this quantitatively expressed?

8. How can you change the extent of a reaction in which a complex is formed or decomposed?

9. When a complex ion takes part in a reaction, how does the rate depend on the nature of the complex?

10. What uses have been found for complexes? Can you justify their study? What natural products contain complex ions?

11.

12.

YOU WILL NEED TO KNOW

Concepts

1. How to write electron configurations and draw orbital diagrams for transition metal atoms and ions — Chapter 6
2. The geometries of atomic orbitals — Chapter 6; and of hybrid atomic orbitals — Chapter 7
3. How to write Lewis structures for molecules and ions — Chapter 7
4. The concept of equilibrium; Le Chatelier's principle — Chapter 13

Math

1. How to calculate K_c; how to use K_c (and K_{sp} and K_a, K_b; see Chapters 16 and 17) to calculate equilibrium concentrations; to calculate the direction in which a system will move to reach equilibrium — Chapters 13, 16, 17

CHAPTER SUMMARY

This chapter offers the opportunity for you to consider the real nature of most species in water solution. In particular, transition metal ions are generally complexed by the solvent or by some other, more strongly basic ligand (basic in the Lewis sense). You have already encountered complexes: recall that the explanation for the acidity of metal ions (Chapter 17) involved aquo complexes. And you may have directly observed how their formation will permit the separation of metal ions in qualitative analysis. Were these reasons not enough justification for studying complexes, we could also point out the existence of many biologically important compounds in which central atoms are bonded to more than their expected share. The energy-storing cytochromes are protein-encapsulated complexes of iron, with the metal ion bonded to six other atoms. Many common drugs, aspirin among them, may derive their potency from their ability to serve as chelating agents.

The very existence of complexes, particulary those of the transition metals, depends on there being low-energy, closely spaced orbitals available for donor pairs of electrons to go into. The d orbitals and the s and p orbitals of adjacent principal energy levels are thus most often used. For most of the transition metals, the d orbitals are at least partially empty and therefore available for bonding.

Crystal field theory is rather successful in predicting how the relative energies of these orbitals on the central atom change with the nature of the ligands. These orbital energy changes in turn provide an explanation for the dependence of properties such as color, behavior in a magnetic field, relative stability (measured, for example by K_c) . . . on the nature of the ligands. In general, for example, the stronger Lewis bases are predicted, and observed, to form the more stable complexes. Likewise, the stronger Lewis acids, those ions of greater charge density, form the more stable complexes.

A relatively simple and generally successful picture of the bonding in complexes is given by valence bond theory. For the complexes we have considered in this chapter, the following correlations can be made: When two bonds are formed by the central atom, that atom employs two equivalent, hybrid atomic orbitals of the type *sp* (such hybrids are directed along a straight line away from the central atom). When four bonds form, hybrids are either sp³ (with tetrahedral geometry) or dsp² (square), depending on what orbitals are available. When six bonds are formed, six equivalent hybrids are derived from two *d*, one *s* and three *p* orbitals (d²sp³ or sp³d²). There are other hybrids that could be described; there are other coordination numbers (numbers of atoms bonded to the central one). The four mentioned are the most frequently encountered. Note that electron pair repulsion ideas would equally as well allow you to predict most of the geometries, knowing simply the coordination number. (See Chapter 7.)

BASIC SKILLS

1. Given the composition of a complex, state its charge.

The principle here is very simple: the charge of a complex is the algebraic sum of that of the central atom and those of the ligands. It is illustrated for complexes of Pt^{2+} in Table 19.1 and the accompanying discussion. Problem 19.1 is entirely analogous. In Problem 19.21, the procedure is reversed; you are asked to determine the charge on the central atom, given that of the complex. Problems 19.2 and 19.22 are related; once you have decided upon the charge of the complex, it is easy to predict its conductivity (cf. Table 19.1).

2. Given the composition of a complex, sketch its geometry, including any geometrical isomers.

In essence, this skill involves three steps.

a) *Determine the coordination number.* This is ordinarily easy; the species $Co(NH_3)_6^{3+}$, $Zn(NH_3)_4^{2+}$, and $Ag(NH_3)_2^{+}$ obviously have coordination numbers of 6, 4, and 2 respectively. If one or more of the ligands is a chelating agent, the coordination number may not be so obvious. Consider, for example, the complexes $Co(NH_3)_4(en)^{3+}$, $Co(NH_3)_2(en)_2^{3+}$, and $Co(en)_3^{3+}$. Since each ethylenediamine molecule bonds to Co^{3+} at two different places, the coordination number in each of these species is 6 (i.e., 4 + 2, 2 + 2(2), 2 × 3).

b) *Relate the coordination number to the geometry.* If the coordination number is 2, the complex is linear; a coordination number of 6 corresponds to an octahedral complex. For coordination number 4, two geometries are possible: tetrahedral or square planar. There is no way to predict from first principles which geometry will prevail with a particular ion.

c) *Having decided upon the geometry, sketch the complex.* This is entirely straightforward for linear and tetrahedral complexes. With square planar complexes you must keep in mind the possibility of geometrical isomerism in complexes of the form $MA_2 B_2$, where A and B are different ligands. Writing all of the isomers for an octahedral complex can be a little tougher. Example 18.1 suggests a logical approach to follow. Note in particular that for any given site in an octahedral complex, there is one site that is trans to it and four that are cis. This remains true regardless of which site you choose as your starting point.

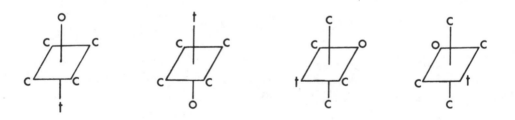

o = original site; c = cis position; t = trans position

Problems 19.8, 19.9, 19.28 and 19.29 illustrate these principles. Where the coordination number is 4, refer back to Table 19.3 to find out whether the complex is likely to be square planar or tetrahedral.

3. Given the composition of a complex ion and its geometry, draw an orbital diagram for the electrons around the central metal atom (valence bond model).

Several such diagrams appear on pp. 517-8; the process used to arrive at them is indicated in Example 19.2. Note that:

- the orbital diagram ordinarily includes only those electrons beyond the nearest noble gas. In the structures on these pages, the 18 argon electrons are not shown.
- the number of orbitals filled by the bonding electrons, contributed by the ligands, is always equal to the coordination number, regardless of the nature of the ligands.

A systematic approach to drawing orbital diagrams might involve the four steps listed below. For simplicity, we assume here that the metal involved is in the first transition series (atomic number 21 – 30), but the same general approach can be extended to other metals.

a) Decide which orbitals are filled by the bonding electrons. This requires only that you know the geometry of the complex.

Coord. no.	Geometry	Hybridization	Orbitals filled (1st transition series)
2	linear	sp	one $4s$, one $4p$
4	square planar	dsp^2	one $3d$, one $4s$, two $4p$
4	tetrahedral	sp^3	one $4s$, three $4p$
6	octahedral	d^2sp^3	two $3d$, one $4s$, three $4p$
		sp^3d^2	one $4s$, three $4p$, two $4d$

For an octahedral complex, assume d^2sp^3 (inner) unless you are told that it is an outer (sp^3d^2) complex.

b) Fill these orbitals with the bonding electrons and leave them there!

c) Decide how many electrons are contributed to the orbital diagram by the central metal ion. To do this, start with the atomic number of the metal, subtract those electrons lost in forming the ion, and finally take away the 18 Ar electrons.

d) If possible, fit the electrons left after (c) into the $3d$ orbitals, following Hund's rule (p. 142). Occasionally, you may find that there are too many electrons to fit into the available $3d$ orbitals. If this happens, put the overflow into the next highest empty orbital.

The application of this approach is shown in the following example.

- -

Draw orbital diagrams for the square planar and tetrahedral complexes of Ni^{2+}. _____

We start by noting that in the square planar complex the hybridization is dsp^2, while in the tetrahedral complex it is sp^3 (step a). These orbitals are then filled by the bonding electrons (step b).

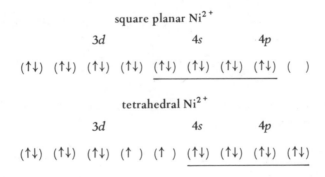

Since the atomic number of Ni is 28, there are 28 electrons in the Ni atom. In the Ni^{2+} ion, there are $28 - 2 = 26$ e^-. Subtracting the 18 argon electrons gives us $26 - 18 = 8$ e^- contributed to the orbital diagram by the Ni^{2+} ion (step c). These are just enough to fill all the available $3d$ orbitals in the square planar complex. In the tetrahedral complex, we distribute the 8 electrons among the five $3d$ orbitals so as to give the maximum number of unpaired electrons (step d).

square planar Ni^{2+}

| 3d | 4s | 4p |

(↑↓) (↑↓) (↑↓) (↑↓) (↑↓) (↑↓) (↑↓) (↑↓) ( )

tetrahedral Ni^{2+}

| 3d | 4s | 4p |

(↑↓) (↑↓) (↑↓) (↑) (↑) (↑↓) (↑↓) (↑↓) (↑↓)

- -

See Problems 19.10 and 19.30.

4. Given the dissociation constant for a complex ion:

a) relate the concentrations of "free" metal ion, complex ion, and ligand.

A typical calculation is given in Example 19.4. See also Problems 19.16 and 19.36. No new principle is involved here. The equilibrium constant for the dissociation of a complex ion is handled in exactly the same way as that for the dissociation of a weak acid (Chapter 17).

b) use the Law of Multiple Equilibria to calculate K for the process of dissolving an insoluble compound in a complexing agent.

This skill is illustrated on p. 527, where it is shown that for the reaction:

$$AgX(s) + 2NH_3(aq) \rightarrow Ag(NH_3)_2{}^+(aq) + X^-(aq); \; K = \frac{K_{sp} \, AgX}{K_d \, Ag(NH_3)_2{}^+}$$

See Problems 19.18 and 19.38, which are entirely analogous to the silver halide-ammonia reaction.

5. Write net ionic equations for reactions involving the formation of complex ions.

Complex ions may be formed by:

a) *direct reaction between metal ion and ligand in solution.* An example would be the reaction that occurs when aqueous ammonia is added to a solution of a Cu^{2+} salt, such as $CuSO_4$. This is described by the equation:

$$Cu^{2+}(aq) + 4NH_3(aq) \rightarrow Cu(NH_3)_4{}^{2+}(aq)$$

A more exact equation would take into account the fact that the Cu^{2+} ion in water is actually in the form of an aquo complex, $Cu(H_2O)_4{}^{2+}$, and that the reaction really occurs by ligand exchange, with NH_3 replacing H_2O:

$$Cu(H_2O)_4{}^{2+}(aq) + 4NH_3(aq) \rightarrow Cu(NH_3)_4{}^{2+}(aq) + 4H_2O$$

However, the simpler equation is commonly used.

b) *reaction between the ligand in solution and an insoluble solid containing the metal ion.* An example is the reaction that occurs when insoluble $Cu(OH)_2$ is treated with aqueous ammonia. The Cu^{2+} ion is brought into solution as the $Cu(NH_3)_4{}^{2+}$ complex:

$$Cu(OH)_2(s) + 4NH_3(aq) \rightarrow Cu(NH_3)_4{}^{2+}(aq) + 2OH^-(aq)$$

Problems 19.17 and 19.37 require that you write net ionic equations similar to these. In addition, Problems 19.19 and 19.39 can perhaps best be answered through net ionic equations, properly interpreted. Note that in certain parts of these problems, two reactions and hence two equations are involved. Thus, addition of OH^- to a solution containing Al^{3+} first precipitates the hydroxide and then further reacts to form the $Al(OH)_6{}^{3-}$ complex. Again (19.37c), addition of NH_3 to silver bromide forms the $Ag(NH_3)_2{}^+$ complex which, like most ammine complexes, decomposes in acid:

$$Ag(NH_3)_2{}^+(aq) + 2H^+(aq) \rightarrow Ag^+(aq) + 2NH_4{}^+(aq)$$

*6. For any octahedral complex, draw diagrams for the "high spin" and "low spin" forms (crystal field model).

The process involved here is indicated in Example 19.3. See Problems 19.12 and 19.32.

Problems

1. Complete the following table.

Metal Ion	Ligands	Formula of Complex
Cu^{2+}	$4\ NH_3$	_____
Cr^{3+}	$5\ NH_3, 1Cl^-$	_____
Co^{2+}	_____	$Co(NH_3)_4Cl_2$
_____	_____	$Co(NH_3)_4Cl_2{}^+$

2. Consider the ions $Co(NH_3)_5Cl^{2+}, Co(NH_3)_4Cl_2{}^+$ and $Co(en)_2Cl_2{}^+$.

    a) What is the coordination number of Co^{3+} in each complex?
    b) What is the geometry of each complex?
    c) Sketch each complex, including geometrical isomers.

3. Consider the "inner" and "outer" octahedral complexes of Co^{2+}.

    a) What orbitals are occupied in the inner complex? the outer complex?
    b) Locate the bonding electrons in orbital diagrams for the two complexes.
    c) How many electrons does Co^{2+} contribute to the orbital diagrams?
    d) Complete the orbital diagrams for the two complexes.

4. The dissociation constant for the $Zn(NH_3)_4^{2+}$ complex is 3×10^{-10}.

   a) What is the ratio $[Zn^{2+}]/[Zn(NH_3)_4^{2+}]$ in 0.10 M NH_3?
   b) Calculate K for the reaction:
      $ZnS(s) + 4NH_3(aq) \rightarrow Zn(NH_3)_4^{2+}(aq) + S^{2-}(aq)$
      (K_{sp} $ZnS = 1 \times 10^{-23}$).

5. Write net ionic equations for:

   a) the reaction of Zn^{2+} with OH^- to form a complex ion.
   b) the reaction of $Zn(OH)_2$ (s) with OH^-.

*6. Draw crystal field diagrams for high spin and low spin octahedral complexes of Co^{3+}.

----

7. Consider the complex formed by Mn^{2+} with three NH_3 molecules and 3 Br^- ions.

   a) Write the formula of the complex ion.
   b) Write the formula of a compound containing this complex ion.
   c) Write the formula of a simple salt whose conductivity in water solution would be similar to that of the compound in (b).
   d) Sketch the geometry of the complex ion, including any isomers.
   e) Draw an orbital diagram for the complex (inner).

8. Consider the complex $Fe(CN)_6^{3-}$, which has a dissociation constant of 1×10^{-31}.

   a) Write a net ionic equation for the formation of the complex from CN^- and metal ion in solution.
   b) Write a net ionic equation for the formation of the complex when an aqueous solution of KCN is added to $Fe(OH)_3$ (s).
   c) Calculate K for the reactions in (a) and (b).
      K_{sp} $Fe(OH)_3 = 5 \times 10^{-38}$.
   d) Would you expect $Fe(OH)_3$ to be appreciably soluble in KCN?

SELF-TEST

True or False

   1. Of the two elements, calcium and copper, the more likely to ( )
form coordination compounds is copper.

2. Of the two ions, CN^- and SO_4^{2-}, the more likely to be found ( ) in metal complexes is the sulfate ion, the weaker base.

3. The sign of the entropy change for the reaction: ( )

$$Ni^{2+}(aq) + dimethylglyoxime \rightarrow chelate(s)$$

is expected to be negative.

4. One would predict that $[Pt(NH_3)_4]Cl_2$ is more soluble in ( ) water than is $[Pt(NH_3)_2Cl_2]$.

5. The smaller the value of the dissociation constant for a ( ) complex, the weaker the coordinate bonds in the complex.

6. The dissociation constant for a complex ordinarily increases ( ) as the strength of the ligands as bases decreases.

7. Any complex ion with a coordination number of four for the ( ) central atom has a tetrahedral structure.

8. Geometric (cis-, trans-) isomerism is not observed in tetra- ( ) hedral complexes.

9. For a complex ion, labile means exactly the opposite of ( ) stable.

10. The color often associated with complexes is best explained ( ) by the approach of valence bond theory.

Multiple Choice

11. In the $NiCl_4^{2-}$ ion, the total number of electrons around ( ) the Ni is
    a) 28   b) 34   c) 36   d) 38

12. The formula $PdCl_2(OH)_2^{2-}$ is known to represent two ( ) different ions. The hybrid orbitals occupied by the bonding electrons are
    a) sp   b) sp^3 c) dsp^2 d) d^2sp^3

13. The hybridization of gold in $Au(NH_3)_2^+$ is: ( )
    a) dsp^2 b) dsp   c) sp^3 d) sp

14. The maximum number of possible geometric isomers for a ( ) complex having sp^3 hybridization would be
    a) 2   b) 3   c) 4   d) none

15. Geometric isomers would be expected for: ( )
    a) $Zn(NH_3)_4^{2+}$ b) $Zn(H_2O)_2(OH)_2$
    c) $Co(NH_3)_3Cl_2Br$ d) $Au(NH_3)_2^+$

16. Complex ions are held together by   ( )
    a) marital bonds   b) municipal bonds
    c) coordinate covalent bonds   d) James bonds

17. Complex ions of coordination number six have a geometric ( )
structure that is
    a) square b) linear   c) tetrahedral   d) octahedral

18. It is known that the sulfhydryl group, $-SH$, forms strong ( )
coordinate bonds to certain heavy metal ions. Which of the following
do you expect to be the best chelating agent for heavy metal ions?
    a) CH_3-SH b) $H-SH$
    c) $CH_3-S-S-CH_3$ d) $HS-CH_2-CH-CH_2-OH$
                                                     SH

19. Which species in the following reaction acts as a Lewis acid? ( )

    $CuSO_4 (s) + 4 NH_3 (aq) \rightarrow Cu(NH_3)_4{}^{2+}(aq) + SO_4{}^{2-}(aq)$

    a) Cu^{2+} b) NH_3 c) $SO_4{}^{2-}$ d) $Cu(NH_3)_4{}^{2+}$

20. The solubility of AgCl in water may be increased by the ( )
addition of
    a) NaCl   b) $AgNO_3$ c) H_2O d) NH_3

21. Of the two complexes, $Cu(H_2O)_4{}^{2+}$ and $Cu(NH_3)_4{}^{2+}$, the ( )
second is the more stable. This means that
    a) $Cu(NH_3)_4{}^{2+}$ would be the stronger acid
    b) ethylenediamine would replace H_2O faster than it would
        replace NH_3
    c) $Cu(NH_3)_4{}^{2+}$ has a smaller dissociation constant
    d) $Cu(NH_3)_4{}^{2+}$ has a larger dissociation constant

22. The electronic structure for the central atom in $Co(en)_2 Cl_2{}^+$ ( )
is:
    a) $(\uparrow\downarrow)\,(\uparrow\downarrow)\,(\uparrow\downarrow)\,(\quad)\,[(\uparrow\downarrow)\quad\quad(\uparrow\downarrow)\quad\quad(\uparrow\downarrow)\,(\uparrow\downarrow)](\quad)$
    b) $(\uparrow\downarrow)\,(\uparrow\downarrow)\,(\uparrow\downarrow)[(\uparrow\downarrow)\,(\uparrow\downarrow)\quad\quad(\uparrow\downarrow)\quad\quad(\uparrow\downarrow)\,(\uparrow\downarrow)\,(\uparrow\downarrow)]$
    c) $(\uparrow\downarrow)\,(\uparrow\downarrow)\,(\uparrow\downarrow)\,(\uparrow\downarrow)\,(\uparrow)\quad\quad[(\uparrow\downarrow)\quad\quad(\uparrow\downarrow)\,(\uparrow\downarrow)\,(\uparrow\downarrow)]$
    d) $(\uparrow\downarrow)\,(\uparrow)\,(\uparrow)\,(\uparrow)\,(\uparrow)\quad\quad[(\uparrow\downarrow)\quad\quad(\uparrow\downarrow)\,(\uparrow\downarrow)\,(\uparrow\downarrow)]$

23. A compound has the empirical formula $CoCl_3 \cdot 4 NH_3$. One ( )
mole of it yields one mole of AgCl on treatment with excess $AgNO_3$.
Ammonia is not removed by treatment with concentrated sulfuric
acid. The formula of the compound is best represented by
    a) $Co(NH_3)_4 Cl_3$ b) $[Co(NH_3)_4] Cl_3$
    c) $[Co(NH_3)_3 Cl_3] NH_3$ d) $[Co(NH_3)_4 Cl_2] Cl$

24. The dissociation constant for the complex ion $Zn(NH_3)_4^{2+}$ is ( )
the equilibrium constant for the reaction represented by the
equation:

a) $Zn^{2+}(aq) + 4 NH_3(aq) \rightleftharpoons Zn(NH_3)_4^{2+}(aq)$
b) $Zn(NH_3)_4^{2+}(aq) + H_2O \rightleftharpoons Zn(NH_3)_3 (H_2O)^{2+}(aq) + NH_3(aq)$
c) $Zn(NH_3)_4^{2+} + 2 e^- \rightleftharpoons Zn(s) + 4 NH_3(aq)$
d) $Zn(NH_3)_4^{2+}(aq) + 4 H_2O \rightleftharpoons Zn(H_2O)_4^{2+}(aq) + 4 NH_3(aq)$

25. Of the following 1.0M solutions, which has the greatest molar ( )
entropy?

a) NaCl   b) $CuCl_2$ c) $AlCl_3$ d) $[Co(NH_3)_5 Cl] Cl_2$

26. To account for the fact that $Fe(NO_3)_3$ dissolves in water to ( )
give an acidic solution, we might write:

a) $H_2O \rightleftharpoons H^+(aq) + OH^-(aq)$
b) $Fe(NO_3)_3(s) \rightleftharpoons Fe^{3+}(aq) + 3 NO_3^-(aq)$
c) $Fe^{3+}(aq) + 3 H_2O \rightleftharpoons Fe(OH)_3(s) + 3 H^+(aq)$
d) $Fe(H_2O)_6^{3+}(aq) \rightleftharpoons Fe(H_2O)_5 (OH)^{2+}(aq) + H^+(aq)$

27. How might you experimentally show that a particular ( )
complex is square?

a) two isomers might be isolated; they would show different
lability
b) x-ray diffraction for the complex in the solid state would
reveal the geometry
c) magnetic measurements might distinguish between dsp^2
and sp^3 orbitals
d) all of the above

28. When $[Ni(NH_3)_4]^{2+}$ is treated with concentrated HCl, two ( )
compounds having the same formula, $Ni(NH_3)_2 Cl_2$, designated I and
II are formed. Compound I can be converted to compound II by
boiling in dilute HCl. A solution of I reacts with oxalic acid,
$H_2 C_2 O_4$, to form $Ni(NH_3)_2 (C_2 O_4)$. Compound II does not react
with oxalic acid. Compound II is:

a) the same as compound I   b) the *cis* isomer
c) the *trans* isomer   d) tetrahedral in shape

SELF-TEST ANSWERS

1. T (Transition metal.)
2. F (The base donates the bonding electrons.)
3. F (Consider that several moles of water are produced. The water is
in the aquo complex, usually indicated merely by (aq).)
4. T (The first gives ions in solution; the second is a nonelectrolyte.)

5. **F**
6. **T**
7. **F** (May also be square.)
8. **T**
9. **F** (Labile refers to rate; stable, to equilibrium position.)
10. **F** (Crystal field or ligand field theory.)
11. **b** (For Ni^{2+} and four pairs of electrons from the four Cl^- ions.)
12. **c**
13. **d**
14. **d** (See question 8 above.)
15. **c**
16. **c**
17. **d** (With six vertices.)
18. **d** (This would permit ring formation. How would you treat heavy metal poisoning?)
19. **a**
20. **d** (To form a complex.)
21. **c**
22. **b** (en is a chelating agent.)
23. **d**
24. **d** (Choice *b* would be only the first step in the overall dissociation that is usually represented by K. Note that you generally omit the water from the equation in *d*.)
25. **c** (Giving 4 moles of ions, compared with 3 for the complex.)
26. **d** (No precipitate forms, as *c* would indicate.)
27. **d**
28. **c**

SELECTED READINGS

Bailar, J. C. Jr., Some Coordination Compounds in Biochemistry, *American Scientist* (1971), pp. 586-592.
   An anecdotal introduction to many interesting and important complexes.
Basolo, F., and Johnson, R. C., *Coordination Chemistry*, New York, W. A. Benjamin, 1964.
   A general extension of the material in this chapter, at about the same level. Molecular oribtal and crystal field theories are emphasized.
Cotton, F. A., Ligand Field Theory, *J. Chem. Ed.* (September 1964), pp. 466-476.
   An extensive, somewhat advanced discussion.
House, J. E. Jr., Substitution Reactions in Metal Complexes, *Chemistry* (1970), pp. 11-14.
   A brief and very readable discussion of several reaction mechanisms. (You might want to first review Chapter 14.)
Pauling, L., *The Nature of the Chemical Bond*, 3rd Ed., Ithaca, N. Y., Cornell, 1960.
   Chapters 5 and 9 discuss the valence bond approach to coordination chemistry. Won't be easy reading.

Perutz, M. F., The Hemoglobin Molecule, *Scientific American* (November 1964), pp. 64-76.

The discussion of this very complicated molecule involving an iron complex includes some background material on x-ray diffraction. A prime example of modern structural chemistry achievements.

Schubert, J., Chelation in Medicine, *Scientific American* (May 1968), pp. 40-50.

Relating physiological properties of a drug to its chemical properties and structure is usually a very difficult and incomplete task. Here's a good beginning.

20

Oxidation and Reduction: Electrochemical Cells

QUESTIONS TO GUIDE YOUR STUDY

1. What reactions have you already encountered that can be classified as oxidation or reduction? What, for example, is a reducing flame? an oxidizing atmosphere?

2. How do you recognize what is oxidized and what is reduced in the equation for a redox reaction?

3. How do you write and interpret redox equations?

4. What occurs in an electrolytic cell? in a voltaic cell? (If you could watch the individual atoms, ions and molecules, what would you expect to see?)

5. How do fuel cells differ from other voltaic cells?

6. What energy effects are associated with reactions in electrochemical cells?

7. What generalizations can be made about what reactions may occur in an electrochemical cell? (How do you predict what is oxidized, what is reduced?) Are there any correlations to be made with the periodic table?

8. How is electrical energy quantitatively related to the masses of reacting species in electrochemical cells?

9. What can be said about the rate of redox reactions?

10. How big, in terms of everyday experience, are the common electrical units: coulomb, ampere, volt, watt? (For example, how large a charge flows through your desk lamp?)

11.

12.

YOU WILL NEED TO KNOW

Concepts

1. How to draw Lewis structures — Chapter 7
2. How to write and interpret balanced net ionic equations — Chapter 16
3. How to interpret free energy changes and the sign of ΔG — Chapter 12

Math

1. How to work problems in stoichiometry — Chapter 3

CHAPTER SUMMARY

The last type of aqueous solution reaction that we discuss in this text is oxidation-reduction, which involves a transfer of electrons from a reducing agent such as Zn, H_2, or I^- to an oxidizing agent such as Zn^{2+}, H^+, or I_2. The species which is oxidized increases in oxidation number ($Zn \rightarrow Zn^{2+}$; O.N., $0 \rightarrow +2$); the species which is reduced decreases in oxidation number ($H^+ \rightarrow H_2$; O.N., $+1 \rightarrow 0$).

Oxidation numbers are assigned using a set of arbitrary rules. Perhaps the most important of these rules tells us that the sum of the oxidation numbers of all the atoms in a species is equal to the charge of that species. This rule can be applied to find oxidation numbers of elements in unfamiliar compounds. For example, for the species $HBrO$ and BrO_3^- we find that the oxidation numbers of Br are:

$$HBrO: +1 + \text{O.N. Br} + (-2) = 0; \text{O.N. Br} = +1$$
$$BrO_3^-: \text{O.N. Br} + 3\,(-2) = -1; \text{O.N. Br} = +5$$

It is possible to balance oxidation-reduction (redox) equations without even considering oxidation numbers by using the half-equation method outlined in the text. One advantage of this method is that it suggests a way of analyzing redox reactions which we will find valuable in Chapter 21. Once an equation is balanced, it can be used to make stoichiometric calculations.

Many familiar redox reactions occur in the world around us. The rusting of iron, the combustion of fuels, and most of the processes involved in human metabolism are redox reactions. All of these reactions and many others that we carry out in the general chemistry laboratory (e.g., the reaction of metals with acids) involve a spontaneous electron transfer from reducing agent to oxidizing agent. Any such reaction can be adapted, at least in principle, to produce electrical energy in a voltaic cell. Most of the cells in current use (dry cell, storage battery) consume chemicals that are too expensive to make them practical, large-scale sources of electrical energy. Fuel cells, in which the chemical energy available from the combustion of fuels is converted directly to electrical energy, would seem to offer a possible solution to the energy crisis. The development of such cells has sufficiently progressed that large-scale power plants are now planned.

Nonspontaneous redox reactions can be carried out by supplying electrical energy in an electrolytic cell. Many important metals (Na, Mg, Al) and industrial chemicals (Cl_2, NaOH) are produced in this way. To calculate the "yield" of products in an electrolytic cell from the quantity of electricity supplied, we use Faraday's Laws. As we might expect, electrochemical processes seldom give a 100% yield of the desired product; side reactions divert an appreciable fraction of the electrons passing through the cell.

BASIC SKILLS

1. Given the formula of an ion or molecule, determine the oxidation number of each atom.

The rules for assigning oxidation numbers are listed in bold type on p. 537 and are applied in Example 20.1. See also Problems 20.1 and 20.21. Problems 20.2 and 20.22 are slightly more subtle but involve the same principles.

You should keep in mind that oxidation numbers are assigned in a quite arbitrary manner and do not have any direct physical meaning. Thus, you should not be disturbed if you find that an atom in a molecule or polyatomic ion has an oxidation number of 0 or -1/2.

2. Given the formulas of products and reactants, balance a redox equation by the half-equation method.

The method is described in some detail on pp. 541-3. Perhaps another example would be helpful.

Balance the equation for the reaction

$$Sn^{2+}(aq) + NO_3^-(aq) \rightarrow Sn^{4+}(aq) + NO_2(g),$$

first in acidic and then in basic solution. _____ ; _____
   The oxidation half-equation is:

$$Sn^{2+}(aq) \rightarrow Sn^{4+}(aq) + 2e^-$$

where the two electrons are added to balance the charges. The reduction half-equation involves the NO_3^- ion:

$$NO_3^-(aq) + 2H^+(aq) + e^- \rightarrow NO_2(g) + H_2O$$

Here, it was necessary to add one H_2O molecule to balance oxygen. This in turn required the addition of $2H^+$ to the left to balance hydrogen. Finally, one electron was added to the left to balance charges.
   To obtain the overall equation in acidic solution, we multiply the second half-equation by 2 and add it to the first. This has the effect of "canceling out" the electrons:

$$Sn^{2+}(aq) \rightarrow Sn^{4+}(aq) + 2e^-$$

$$2NO_3^-(aq) + 4H^+(aq) + 2e^- \rightarrow 2NO_2(g) + 2H_2O$$

$$\overline{Sn^{2+}(aq) + 4H^+(aq) + 2NO_3^-(aq) \rightarrow Sn^{4+}(aq) + 2NO_2(g) + 2H_2O}$$

To balance the equation in basic solution, we add 4 OH^- ions to both sides:

$$Sn^{2+}(aq) + 2NO_3^-(aq) + 4H_2O \rightarrow Sn^{4+}(aq) + 2NO_2(g) + 2H_2O + 4OH^-(aq)$$

which simplifies to:

$$Sn^{2+}(aq) + 2NO_3^-(aq) + 2H_2O \rightarrow Sn^{4+}(aq) + 2NO_2(g) + 4OH^-(aq)$$

Refer to Problems 20.6 – 20.8 and 20.26 – 20.28. A couple of hints:

- O_2 in water solution is ordinarily reduced either to H_2O (acidic solution) or OH^- (basic solution).
- It is quite possible for the same species to be both oxidized and reduced (Problem 20.7e).

3. Given or having derived the balanced equation for a redox reaction, relate the amounts of products and reactants.

See Example 20.2 and Problems 20.9, 20.10, 20.29, and 20.30. Note that no new principles are introduced here; the approach is exactly the same as that first used in Chapter 3 in connection with mass relations in balanced equations. If you have difficulty with part(c) of Problems 20.9 or 20.29, it may be because you have forgotten how to use the Ideal Gas Law (Chapter 5).

4. Relate the amount of electricity passed through an electrolytic cell to the amounts of substances produced or consumed at the electrodes.

Example 20.3 illustrates typical calculations of this kind. Note that you must know the half-equation for the electrode process, or at least the change in oxidation number that is involved. In Example 20.4, the concept of gram equivalent weight in redox reactions is introduced. Note that, in any electrolysis:

$$\text{no. of faradays} = \text{no. of GEW}$$

Of the problems of this type at the end of Chapter 20, Problems 20.13, 20.17, 20.33, and 20.37 are the most straightforward. Problems 20.14, 20.15, 20.34, and 20.35 involve the same principle but require additional conversions to arrive at a final answer.

5. Sketch electrolytic and voltaic cells corresponding to a given redox reaction, label anode and cathode, and trace the flow of current through the cell.

A simple electrolytic cell is shown in Figure 20.2. The Zn-Cu^{2+} salt bridge cell shown in Figure 20.6 is typical of voltaic cells. Note that in both types of cells:

— oxidation occurs at the anode, reduction at the cathode;
— within the cell, anions move to the anode, cations to the cathode;
— outside the cell, electrons move out of the anode and into the cathode.

See Problems 20.18, 20.20, 20.38 and 20.40. Note that in 20.18 and 20.38, it is frequently necessary to use inert electrodes; Pt is a safe choice. Problem 20.40 is particularly instructive because it gives you a chance to show both types of cells in a single sketch.

6. Given the equation for a reduction or oxidation half-reaction, relate the GEW of a species involved in that reaction to its gram formula weight.

The basic relation is:

$$GEW = GFW/n$$

where n is the number of moles of electrons involved per mole of the species. You will recall that the same relation applies to an acid-base reaction, except that there n refers to the number of moles of H^+ (or OH^-) per mole of base (or acid).

- -

What is the GEW of NO_2^- in the half-equation

$$NO_3^-(aq) + 2H^+(aq) + 2e^- \rightarrow NO_2^-(aq) + H_2O? \underline{\hspace{2cm}}$$

Applying the defining equation: $GEW\ NO_2^- = \dfrac{GFW\ NO_2^-}{2} = \dfrac{46.0\ g}{2} = 23.0\ g$

- -

See Problems 20.16 and 20.36 (the half-equations should be balanced first).

Problems

1. Give the oxidation number of each type of atom in

   a) Hg_2^{2+} b) N_2O_5 c) HNO_2 d) CrO_4^{2-} e) $S_2O_3^{2-}$

2. Balance the following redox equation in both acidic and basic solutions.

$$NO_3^-(aq) + I^-(aq) \rightarrow NO(g) + I_2(s)$$

3. Using the balanced equation derived in Problem 2 for acidic solution, calculate:

   a) the no. of moles of H^+ required to form 1.23 moles of I_2.
   b) the no. of grams of NO_3^- required to form 1.68 moles of I_2.
   c) the no. of grams of I^- required to react with 6.00 g of NO_3^-.

4. In the electrolysis of an aqueous solution of NaCl:

$$2Cl^-(aq) + 2H_2O \rightarrow 2OH^-(aq) + Cl_2(g) + H_2(g)$$

   a) How many grams of H_2 are produced by 12,100 coulombs?
   b) How long does it take to form 12.0 g of OH^- using a current of 6.00 amperes?

5. Sketch cells in which the following reactions can occur, label anode and cathode, and indicate the direction of flow of current inside and outside the cell.

a) $Ni(s) + 2H^+(aq) \rightarrow Ni^{2+}(aq) + H_2 (g)$; (voltaic cell)
b) $Zn^{2+}(aq) + 2Cl^-(aq) \rightarrow Zn(s) + Cl_2 (g)$; (electrolytic cell)

*6. *For the half-equation $Cr_2O_7{}^{2-}(aq) + 14H^+(aq) + 6e^- \rightarrow 2Cr^{3+}(aq) + 7H_2O$, what are the gram equivalent weights of $Cr_2O_7{}^{2-}$ and Cr^{3+}?*

— — — — — — — — — — — — —

7. Consider the reaction:

$$Fe^{3+}(aq) + Mn^{2+}(aq) \rightarrow Fe^{2+}(aq) + MnO_4{}^-(aq); \text{ (acidic solution)}$$

a) Balance the equation.
b) Give the oxidation numbers of all atoms in all species.
c) Sketch an electrolytic cell in which the reaction could occur, labeling anode and cathode and indicating the direction of flow of current.
d) How many grams of $MnO_4{}^-$ are formed by 20,200 coulombs?
e) How many grams of Fe^{3+} are required to give 8.20 g of $MnO_4{}^-$?
f) How long does it take to produce 2.51 GEW of $MnO_4{}^-$ using a current of 116 amperes?

SELF-TEST

True or False

1. In a redox reaction, the oxidizing agent gains electrons. ( )

2. The lowest oxidation state of a 5A element is -3. ( )

3. Metals frequently show negative oxidation numbers. ( )

4. A redox equation balanced in acidic solution can be con- ( ) verted to apply in basic solution by adding the proper number of H_2O molecules to both sides.

5. The purpose of the iron screen in the Downs Cell is to ( ) prevent Na^+ and Cl^- ions from coming in contact with each other.

6. Cryolite is added in the Hall process for making aluminum so ( ) that the electrolysis can be carried out at a lower temperature.

7. In the electrolysis of a water solution of NaCl, one mole of ( ) OH^- is produced for every mole of Cl^- consumed.

8. Complexing agents such as CN^- are used in many electro- ( )
plating processes to increase the concentration of metal ions.

9. In the electrolysis of a solution of $Ag(S_2O_3)_2{}^{3-}$, three ( )
faradays of electricity are required to form one mole of silver.

10. In any cell, electrolytic or voltaic, the cathode is the negative ( )
electrode.

Multiple Choice

11. The oxidation number of Mn in the $MnO_4{}^-$ ion is ( )
    a) -2 b) $+6$ c) $+7$ d) $+8$

12. The oxidation number of P in H_3PO_4 is ( )
    a) -3 b) $+1$ c) $+3$ d) $+5$

13. Using Figure 20.1, decide which one of the following is not a ( )
reasonable formula for an oxide of chromium.
    a) Cr_2O b) CrO c) Cr_2O_3 d) CrO_3

14. When the half equation: $I_2 + e^- \rightarrow I^-$ is balanced, the coeffi- ( )
cients of I_2, e^-, and I^- are, respectively,
    a) 1, 1, 1   b) 1, 1, 2   c) 1, 2, 2   d) 1, 0, 2

15. When the half equation: $HSO_3{}^- + H_2O \rightarrow SO_4{}^{2-} + H^+ + e^-$ is ( )
balanced, the coefficients, reading from left to right, are
    a) 1, 1, 1, 1, 1   b) 1, 1, 1, 2, 2
    c) 1, 2, 1, 5, 2   d) 1, 1, 1, 3, 2

16. When the half equations in questions 14 and 15 are ( )
combined, it is necessary to multiply the reduction half equation by
____ and the oxidation half equation by ____ before adding.
    a) 1, 1   b) 1, 2   c) 2, 1   d) 1, 3

17. In the electrolysis of molten magnesium chloride, the most ( )
appropriate equation for the anode reaction would be
    a) $Mg^{2+} + 2e^- \rightarrow Mg(s)$
    b) $2 H_2O + 2e^- \rightarrow H_2(g) + 2 OH^-(aq)$
    c) $2 Cl^- \rightarrow Cl_2(g) + 2e^-$
    d) $MgCl_2(l) \rightarrow Mg(s) + Cl_2(g)$

18. In the electrolysis of Al_2O_3, the ratio of the masses of Al and ( )
O_2 produced per hour is:
    a) less than one   b) 2:3   c) 1:1   d) greater than 1

19. In purifying copper by electrolysis, which electrode should ( )
be made of pure copper?
    a) anode   b) both   c) cathode   d) neither

20. A quantity of 20,000 coulombs is equal to how many ( )
faradays?
    a) 0.021   b) 0.21   c) 1.9×10^9 d) 3×10^{-20}

21. In the formation of chromium metal from Cr^{3+}, the number ( )
of grams of Cr (A.W. = 52) produced by one faraday would be
    a) 17   b) 52   c) 104   d) 156

22. In the reduction of MnO_4^- to Mn^{2+}, the gram equivalent ( )
weight of MnO_4^- is ____ times its gram formula weight.
    a) 5   b) 1   c) $\frac{1}{3}$ d) $\frac{1}{5}$

23. When the $Zn-Cu^{2+}$ cell is used to produce electrical energy: ( )
    a) cations move toward the Zn electrode, anions to the Cu
    b) cations move toward Cu, anions toward Zn
    c) cations and anions move toward Zn
    d) cations and anions move toward Cu

24. When a lead storage battery is charged, lead sulfate ( )
    a) is formed at the cathode
    b) is formed at the anode
    c) is formed at both electrodes
    d) is removed from both electrodes

25. Which one of the following reactions could serve as a source ( )
of energy in a fuel cell?
    a) $H_2O(l) \rightarrow H_2(g) + \frac{1}{2} O_2(g)$
    b) $Zn(s) + Cu^{2+}(aq) \rightarrow Zn^{2+}(aq) + Cu(s)$
    c) $CO_2(g) \rightarrow C(s) + O_2(g)$
    d) $C(s) + O_2(g) \rightarrow CO_2(g)$

SELF-TEST ANSWERS

1. T
2. T (As, for example, in completing an octet.)
3. F (Such numbers are associated with highly electronegative elements.)
4. F (OH^-)
5. F
6. T (The melting point is lower for the solution.)
7. T (Thus maintaining electrical neutrality.)
8. F (To reduce conc.)
9. F (One — Ag goes from +1 to 0.)
10. F
11. c

12. d
13. a   (No +1 state.)
14. c
15. d
16. a
17. c
18. d
19. c   (Cu deposited there.)
20. b
21. a   (1 mole e^- reacts with 1/3 mole Cr^{3+}.)
22. d   (What is the change in oxidation number?)
23. b
24. d   (While in use, i.e., discharging, $PbSO_4$ forms at both.)
25. d   (a and c are non-spontaneous under most conditions; what about b?)

SELECTED READINGS

Anderson, R. C., Combustion and Flame, *J. Chem. Ed.* (May 1967), pp. 248-260.
  Essentially a bibliography for this important class of redox reaction. Better yet, read Faraday's The Chemical History of a Candle.
Brasted, R. C., Nature's Geological Paint Pots and Pigments, *J. Chem. Ed.* (May 1971), pp. 323-324.
  The author writes a column on applied and "relevant" chemistry. This article looks at some redox reactions that are indeed colorful.
Lawrence, R. M. and W. H. Bowman, Electrochemical Cells for Space Power, *J. Chem. Ed.* (June 1971) pp. 359-361.
  A brief survey of some of the cells used for space power.
Weissman, E. Y., Batteries: The Workhorses of Chemical Energy Conversion, *Chemistry* (November 1972), pp. 6-11.
  This is primarily a description of the characteristics of various batteries – their construction, operating reactions, uses, etc.

For practice in balancing redox equations, see the problem manuals listed in the Preface. (Barrow gives programmed exercises for the use of the "oxidation number method.")

Oxidation-Reduction Reactions: Spontaneity and Extent

QUESTIONS TO GUIDE YOUR STUDY

1. How do you know what voltage to apply in carrying out an electrolysis reaction? How do you know what voltage to expect in a voltaic cell?

2. What factors determine the potential associated with any given redox reaction? (Can you relate the potential to properties of atoms such as ionization potential and electronegativity?)

3. How can you predict the spontaneity of any given redox reaction? (How have you been able to predict the spontaneity of other reactions?)

4. What factors determine the extent to which a redox reaction proceeds? How can you change the extent of reaction?

5. What quantitative relationships exist between a cell potential and reaction conditions such as temperature, concentrations and pressure?

6. How might you experimentally show that a given redox reaction is reversible? (Can you think of any common example?)

7. How do you decide what materials may be used for electrodes in voltaic cells; in electrolytic cells?

8. Why does the voltage drop during the use of a voltaic cell? Why do batteries "run down" even when not in use?

9. Are voltaic cells practical major sources of energy? (For example, for lighting and heating a house; for running a car?)

10. To what extent can you convert chemical energy into electrical energy and vice versa?

11.

12.

YOU WILL NEED TO KNOW

Concepts

1. How to recognize and define oxidation and reduction; how to balance redox equations; and other concepts developed in the preceding chapter

2. How to interpret the free energy change for a reaction; how to predict the effects of changes in reaction conditions on the free energy change and on the equilibrium constant — Chapters 12, 13

Math

1. How to use logs and antilogs — Appendix 4

2. How to work stoichiometric problems for redox reactions — Chapter 20

3. How to calculate the free energy change for any reaction; the equilibrium constant for the reaction; and how to quantitatively predict the effect of changes in reaction conditions on ΔG and K — Chapters 12, 13

4. How to calculate K for "multiple equilibria" — see Chapter 17

CHAPTER SUMMARY

From measurements on voltaic cells, it is possible to obtain numbers called *standard potentials* which are a quantitative measure of the tendency of a species to be reduced or oxidized. A large positive value for the standard reduction potential implies a species is easily reduced (e.g., Cl_2 (g) $+ 2$ $e^- \rightarrow 2Cl^-$(aq); S.R.P. = +1.36V). Species which are very difficult to reduce have large negative standard reduction potentials (Al^{3+}(aq) $+ 3$ $e^- \rightarrow$ Al(s); S.R.P. = –1.66V). Standard oxidation potentials, which can be obtained by changing the sign of the potential for the reverse reaction, can be interpreted similarly. Species which are difficult to oxidize, such as Cl^-, have large negative potentials (S.O.P. Cl^- = –1.36V); large positive potentials imply a species which is readily oxidized, such as aluminum (S.O.P. = +1.66V).

The standard voltage, $E°$, corresponding to a particular redox reaction can be obtained by adding the standard potentials for the two half-reactions. If the calculated value of $E°$ is positive, we conclude that the reaction is spontaneous at standard concentrations. Such reactions will take place under ordinary laboratory conditions; alternatively, they can serve as a source of electrical energy in a voltaic cell. In contrast, a reaction with a negative $E°$

value is nonspontaneous at standard concentrations; it can be carried out only by supplying electrical energy. The electrolyses of Al_2O_3 and NaCl, discussed in Chapter 20, are examples of reactions in this category.

The $E°$ value for a reaction can be directly related to the standard free energy change and hence to the equilibrium constant for the reaction. The relation is:

$$\log_{10} K = \frac{nFE°}{2.30\ RT}; \text{ or, at } 25°C, \log_{10} K = \frac{nE°}{0.059},$$

where n is the number of moles of electrons transferred in the reaction. From this equation, we see that a redox reaction which has a positive $E°$ value will have an equilibrium constant greater than 1. If $E°$ is negative, K will be less than 1 and the reaction will be nonspontaneous at standard conditions.

Frequently, we need to know the voltage, E, of a cell when reactants and/or products are present at other than standard conditions (1 atm for gases, 1 M for species in aqueous solution). For the general redox reaction:

$$aA + bB \rightarrow cC + dD$$

we write the Nernst equation:

$$E = E° - \frac{0.059}{n} \log_{10} \frac{(\text{conc } C)^c\ (\text{conc } D)^d}{(\text{conc } A)^a\ (\text{conc } B)^b}, \text{ at } 25°C.$$

This equation tells us that the voltage drops $(E < E°)$ if the concentration of a product is increased (e.g., conc $C > 1$ M); we can increase the voltage of a cell by decreasing the concentration of a product (conc $C < 1$ M) or increasing that of a reactant (conc $A > 1$ M). The Nernst equation is useful for obtaining concentrations of ions in solution from voltage measurements, particularly with components too dilute to be analyzed for by ordinary chemical methods. The pH meter works on this principle; K_{sp} values (Chapter 16) and K_w, K_a and K_b for weak acids and bases (Chapter 17) can also be obtained in this way.

In the last two sections of this chapter, we apply the principles reviewed above to organize the descriptive chemistry of redox reactions in water solution. In particular, we examine some of the more important reactions of strong oxidizing agents (species with large positive reduction potentials). Most of these species fall in one of two categories: they are either highly electronegative nonmetals (F_2, Cl_2, O_2), or oxyanions in which the central atom is in its highest oxidation state (MnO_4^-, $Cr_2O_7^{2-}$, NO_3^-). Perhaps the most important redox reaction, at least from an economic standpoint, is the corrosion of iron and steel. Research in this area indicates that corrosion occurs by an electrochemical mechanism. At an anodic area on the surface of an iron object, Fe atoms are oxidized, first to Fe^{2+} and eventually to a product with the approximate composition $Fe(OH)_3$. At the cathode of the tiny voltaic cell, dissolved oxygen is reduced to H_2O molecules or OH^- ions.

BASIC SKILLS

1. Use standard electrode potentials (Table 21.1) to:

a) compare the relative strengths of different oxidizing agents; different reducing agents.

Referring to Table 21.1, the species in the left column can all, at least in principle, act as oxidizing agents. As one moves down the column, from Li^+ at the top to F_2 at the bottom, the SRP becomes more positive (SRP Li^+ = –3.05 V, F_2 = +2.87 V) and hence oxidizing strength increases. (That is, the species become more easily reduced.)
Species in the right hand column of Table 21.1 are all potential reducing agents. As one moves up this column, from F^- at the bottom to $Li(s)$ at the top, the SOP becomes more positive (SOP F^- = –2.87 V, $Li(s)$ = +3.05V) and hence reducing strength increases.

Which of the following is the strongest oxidizing agent: $MnO_2 (s)$, Cd^{2+}, $I_2 (s)$? _____
Which of the following is the strongest reducing agent: Sn^{2+}, $Ni(s)$, $Zn(s)$?
_____

Of the three oxidizing agents, MnO_2 is closest to the bottom of the left column, has the most positive SRP (+1.23 V), and hence is the strongest oxidizing agent. Of the three reducing agents, $Zn(s)$ is nearest to the top of the right column in Table 21.1, has the most positive SOP (+0.76 V), and hence is the strongest reducing agent.

See Problems 21.4 and 21.24.

b) calculate a cell voltage at standard concentrations.

The relationship here is very simple:

$E°$ = SRP of species which is reduced + SOP of species which is oxidized.

It can be used to calculate the voltage which is produced by the spontaneous reaction going on in a voltaic cell (Example 21.1), or the minimum voltage which must be applied to carry out a nonspontaneous reaction in an electrolytic cell (Example 21.2).
See Problems 21.1, 21.21 and 21.22. Note that in Problem 21.22, you are in effect given $E°$ and one of the potentials (SOP of $Ni(s)$, SRP of Ni^{2+}) and are required to search the table for a species with a potential which will add up to $E°$.

c) calculate $\Delta G°$ for a redox reaction.

The relation here is:

$$\Delta G° \text{ (in cal)} = -23{,}060 \text{ n } E°$$

where $E°$ is the voltage at standard concentrations, as calculated in Skill 1(b), and n is the number of moles of electrons transferred in the reaction. Example 21.4 illustrates calculations of this type. See Problems 21.8 and 21.28. In Problem 21.8, you may have to search the table rather carefully to make sure you find the appropriate half-equations that will add up to the given equation.

d) calculate the equilibrium constant for a redox reaction.

The equation is;

$$\log_{10} K = nE°/0.059; \text{ (at } 25°C)$$

where n has the same meaning as in Skill 1(c). This calculation is carried out in Examples 21.5 and 21.6. In part (b) of these examples, the applications of K are considered. The reasoning and calculations required there are entirely analogous to those involved in gaseous equilibria, Chapter 13, or acid-base equilibria, Chapter 17. Compare, for instance, part (b) of Example 21.6 to part (b) of Example 13.1, p. 354.

Of the problems of this type at the end of Chapter 21, Problems 21.9 and 21.29 are the simplest. In Problems 21.10, 21.11, 21.30 and 21.31, you must first calculate K from $E°$ and then carry out typical equilibrium calculations.

e) decide whether or not a given redox reaction will occur spontaneously at standard concentrations.

The principle here is very simple. If the calculated $E°$ is positive, the reaction is spontaneous. It would, for example, occur if the species were mixed in a beaker or test tube; alternatively, it could serve as a source of energy in a voltaic cell. If the calculated $E°$ is negative, the reaction is nonspontaneous. Work would have to be done to make the reaction go. It could, for example, be made to take place by supplying electrical energy in an electrolytic cell. The following example illustrates the application of this principle in its simplest form.

- -

Which of the following reactions are spontaneous at standard concentrations?

(1) $Zn(s) + Ni^{2+}(aq) \rightarrow Zn^{2+}(aq) + Ni(s)$ _____
(2) $Br_2(l) + 2Cl^-(aq) \rightarrow 2Br^-(aq) + Cl_2(g)$ _____

For reaction
(1): $E° = SRP\ Ni^{2+} + SOP\ Zn(s) = -0.25\ V + 0.76\ V = +0.51\ V$
(2): $E° = SRP\ Br_2(l) + SOP\ Cl^- = +1.07\ V - 1.36\ V = -0.29\ V$

Clearly (1) is spontaneous and could serve as a source of energy in a voltaic cell; (2) is nonspontaneous and would have to be carried out in an electrolytic cell.

- -

See Problems 21.2, 21.5, and 21.25. In the latter two problems, you have to decide upon plausible half-reactions whose standard potentials will add up to a positive value of $E°$.

Frequently the application of this principle is rather more subtle than in the example above. You may be given a series of possible reactants and asked to decide whether or not two of these species will swap electrons in a spontaneous redox reaction (i.e., will yield a positive $E°$). Problems 21.6, 21.7, 21.26, and 21.27 are of this type. Example 21.3, p. 571, illustrates in considerable detail a systematic approach to a problem of this type. "Shortcuts" are often possible. In Example 21.3, having found that the two possible oxidation half reactions both have negative potentials, you could deduce that the reduction half-reaction must have a positive potential if $E°$ is to be positive. This would allow you to dismiss out of hand R_1 and R_3. Two further points which you should keep in mind in working problems of this type are:

- You must combine an oxidation half-reaction with a reduction half-reaction. Students sometimes attempt to combine two oxidations or two reductions, thereby obtaining an absurd answer.
- Sometimes there will be more than one combination that will give a positive $E°$. When this happens, there will be competing spontaneous reactions; the one which occurs most rapidly (not necessarily the one with the most positive $E°$ value) will predominate.

Still another application of this principle requires that you decide what happens when a given solution is electrolyzed, using the minimum voltage. Here, you can assume that the reaction which has the smallest negative voltage will occur. Problems 21.3 and 21.23 are of this type. In working these problems, don't overlook the possibility of producing H_2 or O_2 from the water present. The appropriate half-reactions are:

$$2H_2O + 2e^- \rightarrow H_2\ (g) + 2OH^-(aq); \text{ SRP} = -0.83 \text{ V}$$

$$2H_2O \rightarrow O_2\ (g) + 4H^+(aq) + 4e^-; \text{ SOP} = -1.23 \text{ V}$$

2. Use the Nernst equation (Equation 21.11) to calculate:

    a) the voltage of a cell, given $E°$ and the concentrations of all species.

A typical calculation of this sort is shown in Example 21.7. Again, in Example 21.9, the Nernst equation is applied, this time to determine the effect of concentration upon the potential of a half cell. See also Problems 21.13 and 21.33.

    b) the concentration of one species, given those of all other species, E, and $E°$.

- -

To determine the concentration of H^+ in a solution, a student measures the voltage of a cell in which the reaction is:

$$Zn(s) + 2H^+(aq) \rightarrow Zn^{2+}(aq, 1\text{ M}) + H_2\ (g, 1 \text{ atm})$$

and finds it to be 0.17 volt. Given that $E°$ for this cell is 0.76 volt, what is the concentration of H^+? _____

Applying the Nernst equation:

$$E = E° - \frac{0.059}{n} \log_{10} \frac{(\text{conc. } Zn^{2+})\ (P_{H_2})}{(\text{conc. } H^+)^2}$$

and, substituting for E, $E°$, conc. Zn^{2+}, P_{H_2}, and n:

$$0.17 = 0.76 - \frac{0.059}{2} \log_{10} \frac{1}{(\text{conc. } H^+)^2}$$

Realizing that $\log 1/x^2 = -\log x^2 = -2 \log x$:

$$0.17 = 0.76 - \frac{0.059}{2} (-2 \log_{10} \text{ conc. } H^+) = 0.76 + 0.059 \log_{10} \text{ conc. } H^+$$

Solving:

$$\log_{10} \text{ conc. } H^+ = \frac{0.17 - 0.76}{0.059} = \frac{-0.59}{0.059} = -10; \text{ conc. } H^+ = 1 \times 10^{-10}.$$

- -

Problem 21.14 is entirely analogous to the example just worked. Problem 21.34 is similar, except that you first have to decide upon the direction in which the cell reaction proceeds.

By a slight extension of the approach described here, cell voltages can be combined with measured concentrations to determine equilibrium constants for various kinds of reactions. Example 21.8 shows how solubility products can be determined with the aid of the Nernst equation; see also Problem 21.15. Problem 21.35 is the most difficult of this type; from the statement of the problem you have to deduce the value of $E°$ and the concentration of $Ag(NH_3)_2{}^+$.

*3. Given a balanced redox equation and titration data for the corresponding reaction, calculate the concentration of one of the reactant species.

Refer to Example 21.10, p. 590. No new principle is introduced here. The approach followed is entirely analogous to that used with precipitation reactions (compare Example 16.7, p. 441) and acid-base reactions (compare Example 18.4, p. 483).
Problems 21.20 and 21.40 are entirely analogous to Example 21.10.

Problems

1. Using Table 21.1, p. 568:

   a) arrange the following oxidizing agents in order of increasing strength: Co^{3+}, Ba^{2+}, Sn^{4+}, Fe^{3+}.

   b) calculate $E°$ for a cell in which the reaction is:
   $$2Co^{3+}(aq) + Sn^{2+}(aq) \rightarrow 2Co^{2+}(aq) + Sn^{4+}(aq)$$

   c) calculate $\Delta G°$ for the reaction in (b).

   d) calculate K for the reaction in (b).

   e) Which of the following reactions could serve as a source of energy in a voltaic cell? Which would have to be carried out in an electrolytic cell?
   $$Zn(s) + Fe^{2+}(aq, 1\ M) \rightarrow Zn^{2+}(aq, 1\ M) + Fe(s)$$
   $$H_2 S(g, 1\ atm) \rightarrow H_2 (g, 1\ atm) + S(s)$$
   $$O_2 (g, 1\ atm) + 4H^+(aq, 1\ M) + 4I^-(aq, 1\ M) \rightarrow 2H_2 O + 2I_2 (s)$$

2. Consider the cell:

   $$2Ag^+(aq) + Cu(s) \rightarrow 2Ag(s) + Cu^{2+}(aq); E° = +0.46\ V$$

   a) Calculate E when conc. Ag^+ = conc. Cu^{2+} = 0.10 M.

   b) Calculate conc. Cu^{2+} when E = 0.30 V and conc. Ag^+ = 1.0×10^{-4}M.

*3. Consider the reaction:

$$2MnO_4^-(aq) + 10\,I^-(aq) + 16H^+(aq) \rightarrow 2Mn^{2+}(aq) + 5I_2\,(aq) + 8H_2O$$

If 26.2 ml of 0.100 M KMnO$_4$ is required to react with 20.0 ml of a solution of I$^-$, what is the concentration of I$^-$?

---

4. Consider the following half-equations:

$$Ni^{2+}(aq) + 2e^- \rightarrow Ni(s); \quad SRP = -0.25\ V$$
$$H_2\,(g) \rightarrow 2H^+(aq) + 2e^-; \quad SOP = 0.00\ V$$
$$Cu^{2+}(aq) + 2e^- \rightarrow Cu(s); \quad SRP = +0.34\ V$$
$$Ag(s) \rightarrow Ag^+(aq) + e^-; \quad SOP = -0.80\ V$$

a) Write balanced equations for each of the four possible reactions that are obtained by combining these half-equations.
b) Which of the reactions in (a) are spontaneous at standard concentrations?
c) Calculate $\Delta G°$ and K for each of the nonspontaneous reactions in (a).
d) Which of the species listed would react spontaneously with O$_2$ at standard concentrations in acidic solution? with H$_2$?

5. For the reaction

$$MnO_2\,(s) + 4H^+(aq) + 2Cl^-(aq) \rightarrow Cl_2\,(g) + 2H_2O + Mn^{2+}(aq):$$

a) Calculate E when PCl_2 = 1 atm, conc. H$^+$ = conc. Cl$^-$ = 4M, conc. Mn^{2+} = 1 M.
b) In hydrochloric acid, conc. H$^+$ = conc. Cl$^-$. What must be the concentration of HCl for E to be zero? (PCl_2 = 1 atm, conc. Mn^{2+} = 1 M)

SELF-TEST

True or False

1. The strongest oxidizing agents have the largest, most positive ( )
SRP.

2. The standard reduction potentials of F$_2$ and Ag$^+$ are +2.87V ( )
and +0.80V respectively. We conclude that F$^-$ is a better reducing agent than Ag metal.

3. For a certain mixture, positive $E°$ values are calculated for ( )
two different redox reactions. We can be confident that the one with
the higher $E°$ value will occur first.

4. The voltage of a cell in which the reaction: $Cu(s) + 2\ Ag^+(aq)$ ( )
$\rightarrow Cu^{2+}(aq) + 2\ Ag(s)$ occurs will be, at standard concentrations:
S.O.P. Cu + 2 X S.R.P. Ag^+

5. Reactions which are readily reversed by a small change in ( )
concentration are ones in which $E°$ is close to zero.

6. For the cell referred to in Question 4, increasing the concen- ( )
tration of Ag^+ by a factor of 10 will increase the voltage by +0.06V.

7. The oxidizing strength of oxyanions is ordinarily greatest at ( )
low pH.

8. The stable species of the element nitrogen in the +3 ( )
oxidation state in acidic solution is HNO_3.

9. The phrase "cathodic protection" refers to the common ( )
practice of enclosing fragile metal electrodes in plexiglass to prevent
them from being broken.

Multiple Choice

10. Referring to Table 21.1, if the standard reduction potential ( )
of Ni^{2+} were set at 0.00V, that of Mg^{2+} would be
 a) −2.12V   b) +2.12V   c) −2.62V   d) +2.62V

11. The standard reduction potentials of Cl_2 and Cu^{2+} are ( )
+1.36V and +0.34V, respectively. The $E°$ value for the reaction:
$Cu^{2+}(aq) + 2\ Cl^-(aq) \rightarrow Cu(s) + Cl_2(g)$ is
 a) −2.38V   b) −1.70V   c) −1.02V   d) +1.70V

12. The $E°$ values for the following reactions are known to be ( )
positive:

$$A(s) + B^{2+}(aq) \rightarrow A^{2+}(aq) + B(s)$$
$$A(s) + C^{2+}(aq) \rightarrow A^{2+}(aq) + C(s)$$

At standard concentrations, the reaction between B^{2+} and C:
 a) is spontaneous   b) is nonspontaneous
 c) is at equilibrium   d) cannot say

13. Given the following standard reduction potentials: ( )

$$Mn^{2+}(aq) + 2\ e^- \rightarrow Mn(s) \qquad\qquad -1.18\ \text{Volts}$$
$$2\ H_2O + 2\ e^- \rightarrow H_2\ (g) + 2\ OH^-(aq) \qquad -0.83$$
$$I_2\ (s) + 2\ e^- \rightarrow 2\ I^-(aq) \qquad\qquad +0.53$$
$$O_2\ (g) + 4\ H^+(aq) + 4\ e^- \rightarrow 2\ H_2O \qquad +1.23$$

we would predict that the electrolysis of a water solution of MnI_2 would probably produce:

     a) Mn, I_2 b) Mn, O_2 c) H_2, I_2 d) H_2, O_2

14. For a certain redox reaction, E° is positive. This means that: ( )
     a) ΔG° is positive, K is greater than 1
     b) ΔG° is positive, K is less than 1
     c) ΔG° is negative, K is greater than 1
     d) ΔG° is negative, K is less than 1

15. For the reaction: $4\ Al(s) + 3\ O_2\ (g) + 6\ H_2O \rightarrow 4\ Al(OH)_3\ (s)$ ( )
n in the equation: $\Delta G^\circ = -nFE^\circ$ is
     a) 1   b) 2   c) 3   d) 12

16. For the redox reaction: $A(s) + B^{2+}(aq) \rightleftharpoons A^{2+}(aq) + B(s)$, ( )
$K = 10$. When the concentrations of B^{2+} and A^{2+} are 0.5 M and 0.1 M, respectively:
     a) the forward reaction is spontaneous
     b) the system is at equilibrium
     c) the reverse reaction is spontaneous
     d) cannot say

17. It is possible to increase the voltage of a cell in which the ( )
reaction is

$$Zn(s) + Cu^{2+}(aq) \rightarrow Zn^{2+}(aq) + Cu(s)$$

by increasing the
     a) concentration of Zn^{2+}
     b) concentration of Cu^{2+}
     c) size of the Zn electrode
     d) size of the Cu electrode

18. Which one of the following equilibrium constants would be ( )
most difficult to obtain from cell measurements?

     a) K_b for NH_3 b) K_c for $H_2\ (g) + Cl_2\ (g) \rightarrow 2\ HCl(g)$
     c) K_d for $Cu(NH_3)_4^{2+}$ d) K_w for water

19. A solution of NaOH saturated with Cl_2 at room temperature ( )
will contain appreciable concentrations of all but one of the following species. Indicate the exception.
     a) Cl^- b) OH^- c) ClO^- d) ClO_4^-

20. Which one of the following metals reacts with HNO_3 but not ( )
with dilute HCl?

    a) Pt   b) Mg   c) Na   d) Cu

21. Which one of the following metals reacts with dilute HCl but ( )
not with water?

    a) Ag   b) Na   c) Ni   d) Ca

22. In order to convert $CrO_4{}^{2-}$ to $Cr_2O_7{}^{2-}$, one would add ( )

    a) water   b) an acid
    c) an oxidizing agent   d) a reducing agent

23. Corrosion ordinarily occurs more readily in sea water than in ( )
fresh water because:

    a) Na^+ ions in the sea water attack iron
    b) Cl^- ions in the sea water attack iron
    c) sea water is a better electrical conductor
    d) O_2 is more soluble in sea water

24. "Aqua regia," a mixture of conc. HCl and conc. HNO_3, ( )
dissolves certain metals and metal sulfides which fail to dissolve in
conc. HNO_3. The main function of the HCl is to

    a) increase the conc. H^+
    b) furnish Cl_2, a better oxidizing agent than HNO_3
    c) furnish Cl^-, which acts as a complexing agent
    d) convert the metal to gold, which is soluble in HNO_3

SELF-TEST ANSWERS

    1. T
    2. F   (The reverse is true.)
    3. F   (Will depend on relative rates.)
    4. F
    5. T
    6. T   (Write the Nernst equation.)
    7. T   (High $[H^+]$ favors their reduction.)
    8. F   (HNO_2. Why not $NO_2{}^-$?)
    9. F
  10. a   (Their relative tendencies for reduction remain unchanged.)
  11. c
  12. d
  13. c
  14. c
  15. d
  16. a   (Compare concentration ratio to K.)

17. b   (Think like LeChatelier.)
18. b   (Involves three gases.)
19. d   (Halogens tend to disproportionate in basic solution.)
20. d   (Determine E° for the possible reactions.)
21. c
22. b   (Can you write the equation? What are the oxidation states?)
23. c
24. c

SELECTED READINGS

Taube, H., Mechanisms of Oxidation-Reduction Reactions, *J. Chem. Ed.* (July 1968), pp. 452-461.
   A rather advanced discussion of the mechanisms of redox reactions involving transition metal complexes. (A supplement to the article by House, cited in Chapter 19).

For relating E° to reaction spontaneity and free energy change, see the readings of Chapter 12.

22

Nuclear Reactions

QUESTIONS TO GUIDE YOUR STUDY

1. How are nuclear reactions different from "ordinary" chemical reactions? How would you experimentally recognize that a particular reaction was a nuclear reaction?

2. What factors determine whether a particular atom is radioactive? Are there correlations that can be made with the periodic table; with nuclear composition?

3. What are the properties of the various kinds of radiation? How, for example, do they interact with matter? (What chemical reactions occur as a result of interaction with biological systems?)

4. How do reaction conditions such as temperature, pressure and concentration affect the nature of a nuclear reaction?

5. How would you experimentally determine the rate of a nuclear reaction? What can you say about the rate law and reaction mechanism for a given nuclear reaction?

6. What can you say about the spontaneity and extent of a nuclear reaction?

7. What is the difference between "natural" and "artificial" radioactivity; between fission and fusion?

8. What are some of the *chemical* applications of nuclear reactions?

9. What energy effects are associated with nuclear reactions? Can you predict whether a given nuclear reaction will be exothermic or endothermic?

10. What reactions occur in a nuclear power plant? What are some of the advantages and disadvantages of nuclear power? (How, for example, does the cost compare with that of conventional power from fossil fuels?)

11.

12.

YOU WILL NEED TO KNOW

Concepts

1. How to symbolize (and describe) nuclear composition; i.e., how to use the notation that shows the atomic number and the mass number — Chapter 2

2. How to interpret or describe a reaction mechanism — Chapter 14

3. The general lay-out of the common long form of the periodic table — Chapter 6

Math

1. How to work problems involving the first order rate law — Chapter 14

CHAPTER SUMMARY

In this chapter, we considered three different kinds of spontaneous nuclear reactions.

1) *Radioactive decay,* in which an unstable nucleus decomposes, emitting:

$$\text{an alpha particle: } ^{238}_{92}U \rightarrow ^{4}_{2}He + ^{234}_{90}Th$$

$$\text{a beta particle: } ^{234}_{90}Th \rightarrow ^{0}_{-1}e + ^{234}_{91}Pa$$

$$\text{or a positron: } ^{30}_{15}P \rightarrow ^{0}_{1}e + ^{30}_{14}Si$$

and producing a new, more stable nucleus. Several steps may be required to form a nonradioactive nucleus. The decomposition of uranium-238 passes through 14 intermediates (8 α emissions, 6 β) yielding, as a final product, a stable isotope of lead, $^{206}_{82}Pb$.

A few radioactive isotopes, mostly those of the heavy elements, occur in nature. Many others have been produced in the laboratory by bombardment reactions using positively charged particles (protons, deuterons, α-particles), neutrons, or high energy radiation (X-rays, gamma radiation). An important accomplishment in this area has been the synthesis of at least one isotope each of elements of atomic number 93 to 105, the trans-uranium elements.

Radioactive decay follows the first order rate law (Chapter 14). Rates of different decay processes are often compared by citing half-lives, which can vary from a millisecond to many billions of years. Methods based on the decay of naturally occurring radioactive isotopes have been worked out to determine the age of rocks, both lunar and terrestrial, and carbon-containing artifacts.

2) *Fission*, in which a heavy nucleus, commonly $^{235}_{92}U$ or $^{239}_{94}Pu$, splits under neutron bombardment to give two lighter isotopes. A typical reaction is:

$$^{235}_{92}U + ^{1}_{0}n \rightarrow ^{90}_{37}Rb + ^{144}_{55}Cs + 2^{1}_{0}n$$

Fission ordinarily produces an excess of neutrons; provided a certain critical mass of fissionable material is present, a chain reaction can result.

3) *Fusion*, in which two light nuclei combine, e.g.,

$$^{2}_{1}H + ^{2}_{1}H \rightarrow ^{4}_{2}He$$

Processes of this type, unlike other types of nuclear reactions, have large activation energies. They occur at reasonable rates only at very high temperatures. As of this writing, the only feasible way to achieve and maintain these temperatures is by means of a fission reaction.

All spontaneous nuclear reactions evolve large amounts of energy, most of it in the form of heat. The quantity of energy given off per unit mass of reactant increases in the order: radioactive decay $<<$ fission $<$ fusion. The energy change can be accounted for quantitatively by the Einstein relation:

$$\Delta E \text{ (in ergs)} = 9.00 \times 10^{20} \times \Delta m; \Delta E \text{ (in kcal)} = 2.15 \times 10^{10} \times \Delta m$$

Here, unlike ordinary chemical reactions, there is a detectable difference in mass between products and reactants. The enormous amounts of energy evolved in fission and fusion reflect the fact that the binding energy per nucleon is a maximum for isotopes of intermediate mass (Figure 22.6, p. 613). Since the plot of binding energy per nucleon vs. mass number rises very steeply near the origin and falls off more gradually at high mass numbers, considerably more energy is given off in fusion than in fission.

The reactions discussed in this chapter, none of which were known or even suspected a century ago, have had a profound effect upon our lives and our environment. A generation, born since the holocausts of Hiroshima and Nagasaki, has lived with the constant threat of a nuclear disaster that could destroy life on earth. Now, facing an energy crisis, we are becoming aware of the promise of nuclear reactions, particularly those of the fusion type, to supplement and eventually replace our dwindling supply of fossil fuels.

BASIC SKILLS

1. Write a balanced equation for a nuclear reaction, given the identities of all but one of the reactants and products.

The basic principle here is that both the total mass number (superscript at upper left) and the total nuclear charge (subscript at lower left) must "balance," i.e., must have the same value on both sides of the equation.

- -

Complete the following nuclear equations.

a) $^{90}_{38}$ Sr $\rightarrow$ $^{0}_{-1}$e + _____

b) $^{27}_{13}$ Al + _____ $\rightarrow$ $^{30}_{15}$ P + $^{1}_{0}$ n

c) $^{249}_{98}$ Cf + $^{12}_{6}$ C $\rightarrow$ $^{257}_{104}$ Ru + _____ $^{1}_{0}$ n

In (a), the mass number of the isotope produced must be 90; its nuclear charge must be 39 so that the total on each side of the equation will be 38. Referring to the Periodic Table, we see that the product is an isotope of yttrium, $^{90}_{39}$ Y.

In (b), the mass number of the missing reactant must be: 30 + 1 –27 = 4; its nuclear charge must be: 15 – 13 = 2. Clearly, the reactant is $^{4}_{2}$ He.

In order to balance mass numbers in (c), we need four neutrons so as to obtain a total mass number of 261 on both sides.

- -

Problems 22.3, 22.5, 22.6, 22.22, 22.24, 22.25, and 22.26 further illustrate this simple principle. Problem 22.7 is perhaps the most difficult of this type but is readily solved when you realize that the entire mass number change must be accounted for by the α-particles given off, since a β-particle has a mass number of zero.

2. Use the first order rate law and the expression for the half-life to relate the amount of a radioactive species to elapsed time.

The pertinent equations here are those discussed in Chapter 14 in connection with first order reactions:

$$\log_{10} \frac{X_0}{X} = \frac{kt}{2.30}; \quad t_{1/2} = \frac{0.693}{k}$$

Examples 22.1, 22.2, and 22.3 illustrate the application of these equations to radioactive processes. Problems 22.8 – 22.11 and 22.27 – 22.30 can be worked the same way. In Problem 22.28, note that a plot of log X vs. t should give a straight line for a first order reaction. You should realize that no new concepts are introduced here; very similar calculations were carried out in the problems at the end of Chapter 14.

3. Given a table of nuclear masses such as Table 22.4, calculate for a given nucleus its mass decrement (amu), its binding energy (MeV), and the binding energy per nucleon.

- -

What is the mass decrement of $^{28}_{13}$ Al?_____the binding energy?_____the binding energy per nucleon? _____

The mass decrement is defined as the difference between the mass of the nucleus and that of the individual protons and neutrons that make it up. From Table 22.4:

mass 13 protons = 13(1.00728) amu

mass 15 neutrons = 15(1.00867) amu

= 28.22469 amu

mass $^{28}_{13}$ A1 nucleus =   27.97477 amu; mass decrement = 0.24992 amu

To obtain the binding energy, we use the conversion factor 931 MeV = 1 amu

$$\text{binding energy} = 0.24992 \text{ amu} \times \frac{931 \text{ MeV}}{1 \text{ amu}} = 233 \text{ MeV}$$

Since there are 28 nuclear particles, the binding energy per nucleon is:

$$233 \text{ MeV}/28 = 8.32 \text{ MeV}$$

- -

See Problems 22.17 and 22.36.

4. Using Table 22.4, calculate Δm for a nuclear reaction and relate it to the energy change, ΔE.

The change in mass is readily obtained by the same calculation used for the mass decrement in Skill 3. Note that the use of Table 22.4 gives Δm directly in amu per atom of reactant. Thus: $^{239}_{94}$ Pu $\rightarrow$ $^{235}_{92}$ U + $^4_2$He; Δm per atom Pu = 234.9934 + 4.0015 – 239.0006 = –0.0057 amu. The number obtained this way also represents Δm in grams per mole of reactant. That is, for the reaction just cited, Δm per mole of Pu is –0.0057 g.

Once the value of Δm has been obtained, ΔE is readily calculated, using the conversion factors given in Table 22.3, p. 610. Thus, for the reaction above:

$$\Delta E \text{ per atom Pu} = -0.0057 \text{ amu} \times \frac{931 \text{ MeV}}{1 \text{ amu}} = -5.3 \text{ MeV}$$

$$\Delta E \text{ per mole Pu} = -0.0057 \text{ g} \times \frac{2.15 \times 10^{10} \text{ kcal}}{1g} = -1.2 \times 10^8 \text{ kcal}$$

Other conversions of this type are illustrated in Examples 22.4 and 22.5. Of the problems of this type at the end of Chapter 22, Problems 22.14, 22.16, 22.33, and 22.35 are the simplest (assume that if ΔE is negative, the process is "exothermic" or "spontaneous"). Problems 22.37 and 22.38 are similar but somewhat more tedious. Problems 22.18 and 22.19 illustrate some "relevant" applications of mass-energy conversions.

*5. Predict whether an unstable isotope will decay by electron or positron emission.

Isotopes in which the mass number is "too high," i.e., higher than that in the stable isotopes of the same element, are expected to decay by electron emission. An example is $^{90}_{38}$ Sr, where the mass number is higher than the average atomic mass (87.6). If the mass number is "too low" (e.g., $^{84}_{38}$ Sr), we predict positron emission.

$$^{90}_{38} \text{ Sr} \rightarrow {}^{0}_{-1}\text{e} + {}^{90}_{39} \text{ Y}$$

$$^{84}_{38} \text{ Sr} \rightarrow {}^{0}_{1}\text{e} + {}^{84}_{37} \text{ Rb}$$

See Problems 22.12 and 22.31.

Problems

1. Complete the following nuclear equations:

a) $^{230}_{90}$ Th $\rightarrow {}^{4}_{2}$ He + _____

b) $^{214}_{82}$ Pb $\rightarrow {}^{0}_{-1}$ e + _____

c) $^{40}_{19}$ K + _____ $\rightarrow {}^{37}_{17}$ Cl + ${}^{4}_{2}$ He

2. The half-life of a certain radioactive isotope is 10.0 hours.

a) What is the rate constant for the decay of this isotope?
b) If we start with 1.00 g, how much will be left after 4.0 hours?
c) How long will it take for the amount to drop from 1.00 g to 0.360 g?

3. For the $^{19}_{9}$ F nucleus, calculate

a) the mass decrement in amu
b) the binding energy in MeV
c) the binding energy per nucleon

4. Consider the nuclear reaction: $^{222}_{86}Rn \rightarrow ^{218}_{84}Po + ^{4}_{2}He$. Calculate:

    a) Δm in amu per atom of Rn
    b) Δm in grams per mole of Rn
    c) Δm in grams per gram of Rn
    d) ΔE in MeV per atom of Rn
    e) ΔE in kcal per gram of Rn

*5. *Predict whether the following isotopes will decay by electron or positron emission.*

    a) $^{12}_{5}B$ b) $^{11}_{6}C$ c) $^{14}_{6}C$

– – – – – – – – – – – –

6. Consider the decay of $^{210}_{84}Po$, which gives off an α-particle.

    a) Write a nuclear equation for the decay.
    b) Calculate Δm per atom of Po decaying.
    c) Calculate ΔE (kcal) per mole of Po decaying.
    d) If the half-life for the decay process is 140 days, what is the rate constant?
    e) How long will it take for 90.0% of a sample of Po to be converted to lead?

7. Show by calculation which of the two nuclei, $^{7}_{3}Li$ or $^{6}_{3}Li$, has the larger binding energy per nucleon.

8. If the C-14 content of a piece of organic matter is 0.82 times its original value, how old is the sample?

SELF-TEST

True or False

1. The emission of a β-particle leaves the atomic number   (   )
unchanged but increases the mass number by one unit.

2. The extent of deflection in an electrostatic field is greater for   (   )
a β-particle than for an α-particle.

3. Emission of a positron is equivalent to the conversion of a   (   )
proton to a neutron in the nucleus.

4. The most serious effect of low level radiation on the body is   (   )
that it produces severe skin burns.

5. The longer the half-life of a radioactive isotope, the more ( )
rapidly it decays.

6. The very heavy transuranium elements, atomic number 101 ( )
or greater, are most readily prepared by neutron bombardment.

7. The technique of activation analysis is limited to those ( )
elements that have naturally radioactive isotopes.

8. According to the Einstein relation (Equation 22.19), the ( )
fusion of one gram of deuterium would liberate 9.00×10^{20} ergs or
2.15×10^{10} kcal of energy.

9. Probably the most important hazard associated with a ( )
nuclear power plant is the possibility of a nuclear explosion.

10. A plausible way to achieve the high temperatures required for ( )
nuclear fusion is to operate the process in the upper atmosphere.
(Recall the temperature profile of the atmosphere described in
Chapter 15.)

Multiple Choice

11. The emission of an α-particle lowers the atomic number by ( )
_____ and the mass number by _____ respectively.
    a) 1, 1   b) 1, 2   c) 2, 2   d) 2, 4

12. Nuclear reactions differ from ordinary chemical reactions in ( )
all but one of the following ways. Indicate the exception.
    a) The energy evolved per gram is much greater for nuclear
        reactions.
    b) Nuclear reactions occur much more rapidly.
    c) New elements are often formed in nuclear reactions.
    d) In nuclear reactions, reactivity is essentially independent
        of the state of chemical combination.

13. Emission of which one of the following leaves both atomic ( )
number and mass number unchanged?
    a) positron   b) neutron   c) α-particle   d) γ-radiation

14. A certain radioactive series starts with $^{235}_{92}U$ and ends with ( )
$^{207}_{82}Pb$. In the overall process, _____ α-particles and _____ β-particles
are emitted.
    a) 8, 6   b) 14, 10   c) 7, 10   d) 7, 4

15. Which one of the following instruments would be least suit- ( )
able for detecting particles given off in radioactive decay?
a) Geiger counter   b) scintillation counter
c) electron microscope   d) cloud chamber

16. In determining the age of organic material, one measures ( )
a) the time required for half of the C-14 in the sample to decay
b) the ratio of C-14 to C-12 in the sample
c) the percentage of carbon in the sample
d) the time required for half the sample to decay

17. The half life of uranium-238, which decomposes to lead-206, ( )
is about 4.5×10^9 years. A rock which contains equal numbers of

grams of these two isotopes would be _____ years old.
a) less than 4.5×10^9 b) 4.5×10^9
c) more than 4.5×10^9 d) cannot say

18. Bombardment of $^{75}_{33}As$ by a deuteron, ^2_1H, forms a proton and ( )
an isotope which has a mass number of ____ and an atomic number
of ____.
a) 73, 32   b) 75, 33   c) 75, 32   d) 76, 33

19. The element silicon has an atomic weight of about 28. The ( )
isotope $^{30}_{14}Si$ would most likely decay by emitting a(n)
a) positron   b) proton   c) electron   d) alpha particle

20. Which one of the following isotopes would be most likely to ( )
undergo fission?
a) $^{14}_6C$ b) $^{59}_{27}Co$ c) $^{239}_{94}Pu$ d) ^2_1H

21. Which of the isotopes listed in Question 20 would you expect ( )
to have the largest binding energy per nucleon?

22. Which of the isotopes listed in Question 20 would be most ( )
likely to undergo fusion?

23. According to the Einstein relation (Equation 22.19), the ( )
energy given off in a nuclear reaction in which the decrease in mass is
2.0 mg would be
a) 1.8×10^{21} ergs   b) 1.5×10^{20} ergs
c) 9.0×10^{20} ergs   d) some other number

24. The masses of ^4_2He, ^6_3Li and $^{10}_5B$ are 4.0015, 6.0135 and ( )
10.0102 amu, respectively. The splitting of a boron-10 nucleus to
helium-4 and lithium-6 would
a) evolve energy   b) absorb energy
c) result in no energy change   d) cannot say

SELF-TEST ANSWERS

1. **F**   (Reverse is true.)
2. **T**   (Much smaller mass.)
3. **T**
4. **F**
5. **F**
6. **F**   (Heavier particles used.)
7. **F**
8. **F**   (Δm must be one gram.)
9. **F**   (Controversial! See Readings and recent press reports.)
10. **F**   (Concentration of high T particles is too low.)
11. **d**
12. **b**   (Some are very slow — consider the range of half-lives.)
13. **d**
14. **d**   (Total mass change is due to seven α-particles.)
15. **c**
16. **b**
17. **c**   (Contains more moles of Pb.)
18. **d**
19. **c**
20. **c**
21. **b**
22. **d**
23. **d**   (1.8×10^{18}. How many kcal is this?)
24. **b**   (Δm is positive.)

SELECTED READINGS

Hammond, R.P., Nuclear Power Risks, *American Scientist* (March-April 1974), pp. 155-160.
   A sane discussion of many controversial questions.
Harvey, B. G., *Nuclear Chemistry*, Englewood Cliffs, N. J., Prentice-Hall, 1965.
   A more general and detailed introduction to nuclear chemistry, about on the same level as the text.
Hershey, J., *Hiroshima*, New York, Knopf, 1946.
   For fear that we may forget the potential misapplications of nuclear reactions . . . .
Wahl, W. H. and Kramer, H. H., Neutron Activation Analysis, *Scientific American* (April 1967), pp. 68-82.
   An important application of nuclear reactions, particularly useful recently in lunar analyses.

Also see the article by Seaborg cited in Chapter 6, and the book by Holdren cited in Chapter 4.

23

An Introduction to Biochemistry

QUESTIONS TO GUIDE YOUR STUDY

1. How would you isolate a sample of an enzyme or a nucleic acid? How would you test for purity? How would you determine the composition? (Are these procedures basically the same as those already encountered?)

2. How do biological molecules, like those of carbohydrates, proteins and fats, compare in molecular dimensions (e.g., volume and molecular weight) to each other and to the species dealt with so far?

3. What experimental methods are there for determining molecular structure (both the sequence of bonded atoms and their three-dimensional arrangement)?

4. What energy is available to an organism for driving its energy-requiring processes? How is it transferred? How is it stored?

5. What range of conditions (T, P, concentration . . .) prevail during ordinary biochemical reactions? What is known about reaction mechanisms?

6. Do proteins (and other complex biological molecules) spontaneously arise from amino acids (and other simpler molecular units)? Is life itself a spontaneous process?

7. Can we begin to explain the properties of living organisms in molecular terms? (Like growth and reproduction; mutations, disease, death.)

8. What kinds of answers can we now give to how life may have begun; how evolution occurs? (What kinds of questions cannot be, or at least have not been, answered?)

9. Why are there relatively *few* kinds of molecules important to life?

10. Can you now begin to account for the large-scale resolution of matter into simple gaseous substances (the atmosphere) solids and liquids (ores in the earth's crust, the oceans) and biological substances (as confined to living organisms) — a resolution that has occurred through eons and continues to occur?

11. Can you relate this resolution, or fractionation, to simple chemical systems amenable to study in the general chemistry laboratory?

12.

13.

YOU WILL NEED TO KNOW

Concepts

The application of chemistry to biological problems will prove more rewarding the more chemistry you know. Consider some of the concepts tied together in this chapter:

- methods of separation, e.g., chromatography — Chapter 1
- molecular weights — Chapter 2
- structural formulas, Lewis formulas — Chapter 7
- hydrogen bonding and the nature of other inter- and intramolecular forces — Chapter 7
- functional groups and their properties ($-NH_2$, $-CO_2H$, amide . . .) — Chapter 8
- polymers and their properties — Chapter 8
- methods and results of structure determination, e.g., x-ray diffraction — Chapter 9
- principles of solubility — Chapter 10
- catalysis — Chapter 14
- acid-base properties, pH — Chapters 17, 18
- oxidation state, oxidation-reduction — Chapters 20, 21

Math

Except for calculations in the problems involving MW, there is no math used in this chapter.

CHAPTER SUMMARY

For all the sweat or tears of joy you may have shed by now, your own composition is still something like 65 per cent water, 15 per cent protein, 15 per cent fat, and less than 1 per cent carbohydrate – the rest being inorganic compounds. Despite the complexities of the thirty-odd per cent organic compounds and the reactions they take part in, we can note here several simplifications.

1. Though debate continues in the background, the nature of life seems to be explicable in terms of known principles of chemistry and physics. (In particular, no new chemical principles are needed in this chapter.) The molecular interpretation of biological structures and processes is already coming of age. For example: reaction pathways, the energy and material flows, have been discovered for most of the reactions involving the major constituents of living organisms.

2. Certain basic features are known to be common to all organisms: molecules with very similar architectures perform very similar functions, whether in one organism or another, plant, animal, or microbe. For example: the hemoglobin of man is built much like that of the horse and that of the gorilla, and all three serve to carry oxygen. There are really only a few kinds of biological molecules, and they are common to all life as we know it. (The usual classification lists carbohydrates, fats, proteins and nucleic acids as major components.)

3. Much of the chemistry of life occurs in dilute aqueous solution (and so rules out the need for studying many kinds of substances incompatible with water), at or near room temperature. The reactions generally, if not always, involve enzyme catalysis, and often oxidation-reduction as well.

4. Many, if not most, reactions are interlocked (coupled) with others. The energy released in one reaction is, at least in part, used to drive another reaction. For all coupled reactions, there is a net decrease in free energy. There is no creation of order in the living organism that doesn't occur simultaneously with a larger disordering of the surroundings.

In addition to being one of the most active research areas in recent years, molecular biology has been perhaps *the* area of cooperation among the various "compartments" of science. Working for perhaps a dozen man-years in elucidating the amino acid sequence for a single small protein, a team of physicists, chemists and biologists have freely shared the tools and theories perfected individually, while training one another in the problems of their new frontier.

What are some of the unresolved or incompletely solved problems?

1. *Molecular structures* are known in three-dimensional detail for only a handful of biological species. We know most other structures only in sketchy outline. All evidence so far indicates that all the biological properties of a molecule are determined by its primary structure (the number and kinds of atoms present and their bonding sequence).

2. *Molecular functions* and reaction mechanisms have been vaguely described for just a few representative compounds. Again, the details need to be discovered before a thorough understanding can be achieved.
3. *Comparative studies* of molecules serving similar functions need to be pursued in order to further describe the mechanism by which evolution occurs, by which only a few types of molecules seem to have survival value. The current work on compounds associated with fossils, as well as the abiological syntheses of compounds associated with living organisms, should continue to shed light on the possible origins of life itself.
4. Humanistic and moral questions need be raised: most biochemists consider as inevitable deliberate *design* of organisms, as well as modification of organisms already alive. All of us need to know where we may be going.

BASIC SKILLS

This chapter is primarily descriptive; very few new concepts are introduced.

1. Given the structural formula for an organic molecule, state whether or not it will show optical isomerism.

At least for our purposes, we can expect to find optical isomers only if there is one or more asymmetric carbon atom in the molecule (i.e., a carbon atom which is bonded to four different groups). Thus, we expect the following molecules to show optical isomerism:

$$\text{Cl}-\overset{\text{H}}{\underset{\text{Br}}{\text{C}}}-\text{OH}, \quad \text{H}_3\text{C}-\overset{\text{H}}{\underset{\text{Cl}}{\text{C}}}-\overset{\text{H}}{\underset{\text{H}}{\text{C}}}-\text{OH}, \quad \text{H}_3\text{C}-\overset{\text{H}}{\underset{\text{OH}}{\text{C}}}-\overset{\text{H}}{\underset{\text{H}}{\text{C}}}-\text{OH}, \quad \text{H}_3\text{C}-\overset{\text{OH}}{\underset{\text{H}}{\text{C}}}-\overset{\text{OH}}{\underset{\text{H}}{\text{C}}}-\text{CH}_3$$

but not the following:

$$\text{Cl}-\overset{\text{H}}{\underset{\text{Cl}}{\text{C}}}-\text{OH}, \quad \text{H}_3\text{C}-\overset{\text{H}}{\underset{\text{Cl}}{\text{C}}}-\text{H}, \quad \text{H}_3\text{C}-\overset{\text{H}}{\underset{\text{H}}{\text{C}}}-\text{CH}_3, \quad \text{H}_3\text{C}-\overset{\text{H}}{\underset{\text{H}}{\text{C}}}-\overset{\text{OH}}{\underset{\text{OH}}{\text{C}}}-\text{CH}_3$$

See Problems 23.3 and 23.18.

2. Given the sequence of amino acids in a polypeptide, write its structural formula.

Write the structural formula of a tripetide in which the three amino acid residues are glycine, alanine, and serine, i.e., "gly-ala-ser". ————

To arrive at the structural formula, we first write down the structures of the three acids in such a way that the –COOH group of one acid is directly opposite the –NH₂ group of the next. Referring to Table 23.1 for the structures of the three acids:

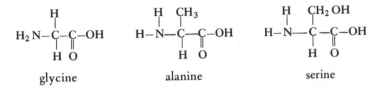

| glycine | alanine | serine |

We then split out molecules of water (OH and H) between adjacent molecules to arrive at:

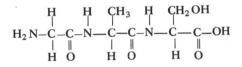

gly-ala-ser

See Problem 23.6.

3. Given the information obtained from hydrolysis of a polypeptide, establish the sequence of amino acid residues.

The following simple example illustrates the principle involved.

Hydrolysis of a polypeptide containing six amino acid residues gives the following information:

— products from complete hydrolysis: Ala(2), Pro(2), Ser(2)
— terminal residues: –NH₂ end = Ala, –COOH end = Ser
— fragments: Ser-Ala-Pro, Pro-Ser-Ala

Indicate the sequence of amino acid residues. ————

To establish the sequence, we might first insert the terminal acids:

Ala ——— ——— ——— ——— Ser

There are two ways in which we might fit in the first fragment:

a) Ala <u>Ser</u> <u>Ala</u> <u>Pro</u> ____ Ser
b) Ala ____ <u>Ser</u> <u>Ala</u> <u>Pro</u> Ser

Clearly, the second fragment could not fit in (a). It could, however, fit in (b):

b) Ala ____ ____ ____ ____ Ser

             Ser   Ala   Pro   (1st fragment)

        Pro   Ser   Ala   (2nd fragment)

So the sequence must be: Ala – Pro – Ser – Ala – Pro – Ser.

Problem 23.7 is somewhat more complicated than this example. One hint: it will be helpful to start with the Pro-Ala-Pro group and consider how the other fragments might overlap it (note that there are only two Pro residues in the molecule).

Problems

1. Which of the following organic compounds would show optical isomerism?

a) $H_3C-CH_2-CH_2-CH_3$ b) $H_3C-CH_2-\overset{\displaystyle OH}{\underset{\displaystyle H}{C}}-CH_3$

c) $HO-CH_2-CH_2-CH_2OH$ d) $HO-CH_2-\overset{\displaystyle OH}{\underset{\displaystyle H}{C}}-CH_2OH$

2. Write structural formulas for the following polypeptides.

a) Val–Ala   b) Lys–Cys–Gly

*3. *Hydrolysis of a certain polypeptide gives the following information:*

  — acid residues present: Ser(2), Pro, Cys(2), Ala(2)
  — end groups: $-NH_2$ end = Ser, $-COOH$ end = Cys
  — fragments: Cys–Pro–Ser, Ala–Cys–Pro, Pro–Ser–Ala
Indicate the sequence of amino acid residues.

----

4. Indicate how each of the following could be converted into a compound with optical isomers by substituting a chlorine for a hydrogen atom.

a)
$$H-\overset{\overset{\displaystyle H}{|}}{\underset{\underset{\displaystyle OH}{|}}{C}}-\overset{\overset{\displaystyle H}{|}}{\underset{\underset{\displaystyle OH}{|}}{C}}-H$$

b) $H_3C-CH_2-CH_2-CH_3$

c)
$$H-\overset{\overset{\displaystyle H}{|}}{\underset{\underset{\displaystyle H}{|}}{C}}-\overset{\overset{\displaystyle OH}{|}}{\underset{\underset{\displaystyle H}{|}}{C}}-CH_3$$

5. Write structural formulas for all the tripeptides that could be formed using 1 serine and 2 glycine molecules.

SELF-TEST

True or False

1. In animals, proteins serve primarily to store energy. ( )

2. Carbohydrate polymers are synthesized only by plants. ( )

3. The net photosynthetic reaction, ( )
$CO_{2\,(g)} + H_2O_{(g)} \rightarrow glucose_{(s)} + O_{2\,(g)}$, involves a reduction of carbon.

4. The α-amino acids might generally be expected to form ( ) chelates with metal atoms.

5. The water solution of any amino acid which contains only ( ) one basic functional group and one acidic group is expected to have a pH of 7.

6. The hydrolysis of proteins (breakage of the peptide links ( ) with insertion of water molecules) will always result in the formation of equal numbers of $-NH_2$ and $-CO_2H$ groups.

7. The formation of an enzyme from amino acids has a positive ( ) free energy change. The fact that the synthesis of enzymes occurs in a living organism is a contradiction of the laws of thermodynamics.

8. Cellular RNA can be thought of as an exact copy of DNA. ( )

Multiple Choice

9. Where rigidity in structure is required (as in an animal's ( )
skeleton), the material is likely to consist of
(a) DNA (b) nucleotides (c) polysaccharrides (d) alkanes

10. Which one of the following formulas would you choose to ( )
represent an average composition of protein?
(a) CH_2O (b) $C_{57}H_{110}O_6$ (c) CH_7NO (d) $C_9H_{13}SN_2O_2$

11. The largest amount of heat would be obtained from the ( )
complete combustion of a gram of:
(a) protein   (b) carbohydrate
(c) fat (e.g., $C_{57}H_{110}O_6$)   (d) glucose

12. The geometry associated with the atoms of the peptide link, ( )

$$\overset{O}{\overset{\|}{C_\alpha-C}}-\overset{H}{\overset{|}{N}}-C_\alpha,$$ is observed to be:

(a) linear   (b) tetrahedral   (c) planar
(d) variable, depending on the rest of the chain

13. The method of choice for determining the molecular weight ( )
of a protein is:
(a) osmotic pressure
(b) gas density
(c) freezing point lowering
(d) direct weighing of a single molecule

14. The best solvent for a polypeptide is likely to be   ( )
(a) CCl_4 (b) CH_3-O-CH_3 (c) H_2O (d) HF

15. To separate a mixture of monosaccharides, you would ( )
probably use a(n):
(a) centrifuge   (b) column chromatograph
(c) mass spectrometer   (d) electrolytic cell

16. In a 0.10 M solution of glycine, $H_2C(NH_2)CO_2H$, at a pH of ( )
10, the most abundant species next to water is
(a) OH^- (b) $H_2C(NH_3)CO_2$
(c) $H_2C(NH_3)CO_2H^+$ (d) $H_2C(NH_2)CO_2^-$

17. The linkage between monomers ("residues") in a nucleic acid ( )
involves
(a) C–N   (b) C–O–C   (c) P–O–C   (d) hydrogen bond

18. The maximum number of different tripeptides that can be ( )
formed from three different amino acids, using one residue of each,
is:

    (a) 2   (b) 3   (c) 4   (d) 6

19. The maximum number of different tripeptides that can be ( )
made from three different amino acids, using any number of residues
of each, is:

    (a) 2   (b) 3   (c) 6   (d) 27

20. A synthetic RNA made up of only the base uracil would ( )
probably cause a cell to produce a polypeptide which contained only
    (a) uracil
    (b) phenylalanine
    (c) three phenylalanines
    (d) any amino acid other than Glu, Gln or Lys

21. As the temperature is increased, the rate of an enzyme- ( )
catalyzed reaction first increases and then decreases. The decrease
can be explained in the following way:
    (a) all reactions achieve a maximum rate at some tempera-
        ture
    (b) the molecules acted on by the catalyst spontaneously
        decompose at the high temperature
    (c) intramolecular forces (e.g., H-bonding) in the enzyme
        molecule begin to break down
    (d) all catalyzed reactions behave in this unaccountable
        manner

22. The rate of an enzyme-catalyzed reaction is generally found ( )
to increase with the concentration of the reactant substrate up to
some maximum rate and then level off. This levelling-off is probably
due to:
    (a) decomposition of enzyme
    (b) a change in enzyme conformation
    (c) molecules of reactant get in the way of each other at high
        concentrations
    (d) all the "active sites" on the enzyme molecules are
        occupied

23. Which one of the following species would you expect *not* to ( )
show optical activity?
    (a) $CH_2(NH_2)CO_2H$ (b) $CH_3CH(NH_2)CO_2H$

           OH O   OH
           | || |
    (c) $CH_3CH{-}CH$ (d) $CH_3CH_2CHCH_3$

24. A change in the composition of which kind of molecule is ( )
most likely to result in a mutation?
    (a) RNA   (b) DNA   (c) carbohydrate   (d) enzyme

25. The "backbone" in a protein molecule can assume only a ( )
limited number of conformations. This is a result of:
    (a) hydrogen bonding
    (b) relative solubilities of attached groups
    (c) the geometry of the peptide link
    (d) all of the above

26. The argument for evolution gains support from which of the ( )
following?
    (a) proteins, nucleic acids and carbohydrates are not
        sufficiently stable to be found in fossils
    (b) the function of a particular enzyme in man is served by
        enzymes of similar structure in other organisms
    (c) the "active site" in an enzyme is the locale for most
        mutations
    (d) DNA is found in all organisms

SELF-TEST ANSWERS

1. F (Fats and glycogen serve in this capacity; proteins, in many
       others.)
2. F
3. T

4. T (Like:
$$\begin{array}{c} \diagdown C{-}O \diagdown \\ | \qquad M \\ {-}C{-}N \diagup \\ | \end{array}$$
)

5. F
6. T
7. F (First, the biosynthesis doesn't begin with free amino acids but
       with complex molecules such as proteins. Second, other reactions
       occur that supply the free energy necessary for the synthesis.)
8. F
9. c (Where the most H-bonding can occur. However, carbohydrates
       serving this function are mainly found in plants; proteins, in
       animals.)
10. d
11. c (Carbon in lowest oxidation state. Contrast C, CO and CO_2 as
        fuels.)

12. c   (Consider the resonance structure: 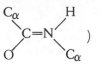)

13. a   (Most sensitive.)

14. c   (A nondestructive solvent for these molecules that contain lots of polar groups.)

15. b

16. d   (What is [OH⁻]?)

16. d   (What is $[OH^-]$?)

17. c

18. d   (Each amino acid residue could be at the $-NH_2$ end or at the $-CO_2H$ end.)

19. d

20. b

21. c

22. d

23. a   (No asymmetric carbon.)

24. b

25. d

26. b

SELECTED READINGS

Calvin, M., *Chemical Evolution,* New York, Oxford University Press, 1969.
   A rather technical and personal discussion that offers much to the reader, whether skimmed or read carefully. Discusses chemical fossils as well as work done in laboratory attempts at synthesizing biological molecules and systems.

Handler, P., Ed., *Biology and the Future of Man,* New York, Oxford University Press, 1970.
   Intended for the well-read "layman"; there is much chemistry here as well as an overview of research in biology in general.

Watson, J. D., *Molecular Biology of the Gene* (2nd edition), Menlo Park, Calif., W. A. Benjamin, 1970.
   A very well-written account of much of modern biochemistry, by one who has pioneered.

Chemical and Engineering News has carried several interesting feature articles relevant to this chapter. Some examples:
   Chemical Origins of Cells (June 22, 1970 & Dec. 6, 1971)
   Some Chemical Glimpses of Evolution (Dec. 11, 1967)
   Life Transcending Physics and Chemistry (Aug. 21, 1967)
   The Synthesis of Living Systems (Aug. 7, 1967)

Scientific American is a very rich resource. Consider recent biochemical articles:
   The Primary Events of Photosynthesis (December 1974)
   Nitrogen Fixation (October 1974)
   The Carbon Chemistry of the Moon (October 1972)
   The Chemical Elements of Life (July 1972)
   The Structure and History of an Ancient Protein (April 1972)

Appendix
Answers To Problems

Chapter 1

1. a) 9.86   b) 6.69   c) 4.2   d) 16.0

2. a) 3.12×10^{-3} kg/cm^3 b) 3.12×10^3 g/liter   c) 3.12×10^3 kg/m^3
   d) 1.13×10^{-1} lb/in^3 e) 195 lb/ft^3

3. a) 1.63×10^3 g   b) 321 cm^3

4. a) $^{\circ}Y = 1.80^{\circ}C + 20^{\circ}$ b) $110^{\circ}Y; 122^{\circ}F; 323^{\circ}K$

5. a) 63 g   b) 5 g   c) 39 g

6. a) 21.26 cm^3 b) 3.74 cm^3 c) 4.19 g/cm^3 d) 0.151 lb/in^3
   e) 74.2 in^3 f) 0.0267 liter

7.

50	100	27	204	300
0	100	-46	308	500
122	212	81	400	572
323	373	300	477	573

8. mp $NH_3 = -78^{\circ}C$; bp $NH_3 = -33^{\circ}C$
   $^{\circ}N = a^{\circ}C + b; 0 = -78a + b; 100 = -33a + b; a = 100/45, b = 7800/45$
   $^{\circ}N = (100^{\circ}C + 7800)/45; 98.6^{\circ}F = 37^{\circ}C; ^{\circ}N = 11500/45 = 256^{\circ}$

9. a) At $100^{\circ}C$, $S(A) = 20 + 40 + 20 = 80; S(B) = 10 + 30 + 40 = 80$

   To dissolve 60 g of A, use $100 \times (60/80) = 75$ g water, which is more than enough to dissolve 20 g of B.

   b) $S(B) = 20 \times (100/75) = 27 = 10 + 0.30t + 0.0040t^2$
   $0.0040t^2 + 0.30t - 17 = 0$; solving by the quadratic formula, $t = 38^{\circ}C$

   c) $S(A) = 20 + 0.40(38) + 0.0020(38)^2 = 38$ g/100 g water

   In 75 g water, $38 \times 0.75 = 28$ g of A remains; 32 g of solid A appears

Chapter 2

1. a) 33, 27, 27; 52, 38, 36   b) $^{23}_{12}Mg; ^{85}_{37}Rb; ^{139}_{57}La$

2. 85.4

3. a) 28.10   b) 76.5% Cl-35

4. 24.3   8.5 $\times$ 10^{-20} 3.5 $\times$ 10^{-21} 2.1 $\times$ 10^3
36.5   59.9   1.64   9.88 $\times$ 10^{23}
78.1   52.0   0.666   4.01 $\times$ 10^{23}

5. 1.08, 2.16, 3.26; in ratio of 1 : 2 : 3

6. 19.0

7. 0.11 cal/g°C

8. 91.0%, 8.7%

9. a) 11.92   b) 11.91   c) 17.87

10. a) 1.06 $\times$ 10^{24} b) 3.76 $\times$ 10^{23} c) 3.39 $\times$ 10^{24}

11. a) 1.05 $\times$ 10^{-22} g   b) 5.32 $\times$ 10^{-9} g   c) 56.0 g

Chapter 3

1. 3.34 $\times$ 10^{-15} 4.01 $\times$ 10^{-14}
37.9

0.200   1.20 $\times$ 10^{23}
8.63 $\times$ 10^{-23} 2.76 $\times$ 10^{-21}

2. a) 42.9% C, 57.1% O   b) 20.2% Al, 79.8% Cl   c) 15.8% Al, 28.1% S, 56.1% O

3. a) Bi_2O_3 b) $Cr_2S_3O_{12}$ c) C_2H_5 d) PbI_2

4. a) $CH_4(g) + 2 O_2(g) \rightarrow CO_2(g) + 2 H_2O(l)$

b) $2 C_6H_6(l) + 15 O_2(g) \rightarrow 12 CO_2(g) + 6 H_2O(l)$

c) $2 CH_3OH(l) + 3 O_2(g) \rightarrow 2 CO_2(g) + 4 H_2O(l)$

5. a) 3.28   b) 195   c) 178   d) 1.96 $\times$ 10^{22}

6. a) Cl_2 b) Cl_2

7. C_4H_{10}

8. 88.9 g; 67.5%

9. a) 1.04 g C, 0.175 g H, 1.38 g O   b) 40.0% C, 6.73% H, 53.1% O

10. a) 71.67%   b) 1.446 g   c) 53.38% Sb, 46.62% Cl   d) $SbCl_3$

11. a) 1.09   b) 2.66 $\times$ 10^{-12} c) 2.66 $\times$ 10^{-12}

12. a) $Al^{3+}(aq) + 3 OH^-(aq) \rightarrow Al(OH)_3(s)$ b) 0.660; 1.64   c) OH^-
d) 0.546; 42.6 g

13. a) no. g Mn in MnO_2 = 0.435 g MnO_2 $\times \dfrac{54.9 \text{ g Mn}}{86.9 \text{ g } MnO_2}$ = 0.275 g Mn

in the second oxide, there must be 0.275 g Mn + 0.107 g O

no. moles Mn = 0.275/54.9 = 5.01 $\times 10^{-3}$; no. moles O = 0.107/16.0 = 6.69 $\times 10^{-3}$

1 Mn : 1.33 O, Mn_3O_4

b) 3 MnO_2 (s) → Mn_3O_4 (s) + O_2 (g)

14. if KCl: 1.00 g KCl $\times \dfrac{143.3 \text{ g AgCl}}{74.6 \text{ g KCl}}$ = 1.92 g AgCl

hence, KCl

if KI : 1.00 g KI $\times \dfrac{234.8 \text{ g AgI}}{166.0 \text{ g KI}}$ = 1.41 g AgI

15. 2 moles $[Co(NH_3)_5 NO_2](NO_3)_2 \triangleq$ 1 mole Co_2O_3

628 g $[Co(NH_3)_5 NO_2](NO_3)_2 \triangleq$ 166 g Co_2O_3

no. g complex = 0.365 g $\times \dfrac{628 \text{ g}}{166 \text{ g}}$ = 1.38 g; % complex = $\dfrac{1.38}{1.50} \times 100$ = 92.1%

Chapter 4

1. a) −12.1 kcal   b) −354 kcal   c) −118 kcal

2. a) −8.4 kcal   b) −85.0 kcal

3. −12 kcal

4. +153 cal

5. −91.8 kcal, +44.8 kcal

6. 0.227; 11 kcal

7. a) C_2H_4 (g) + 3 O_2 (g) → 2 CO_2 (g) + 2 H_2O(l)

b) −337.3 kcal   c) −12.0 kcal   d) 8.30 g

8. a) −79.9 kcal   b) −75 kcal   c) $\Delta H_{vap} CHCl_3$

9. a) −53.7 kcal   b) −67.7 kcal

10. a) −372 kcal   b) −21 kcal

11. From Table 4.1, ΔH for the combustion of one mole of CH_4 by the reaction:

$$CH_4 (g) + 2 O_2 (g) → CO_2 (g) + 2 H_2O(g)$$

is −191.8 kcal. Since there is a 5-fold excess of air, there must be 10 moles of O_2 to start with, and 40 moles of N_2. After reaction, there are 40 moles N_2, 10 − 2 = 8 moles O_2, 1 mole CO_2, and 2 moles H_2O. In grams, there are 1120 g N_2, 256 g O_2, 44.0 g CO_2, 36.0 g H_2O to absorb 191,800 cal of heat.

191,800 = [1120(0.249) + 256(0.209) + 44.0(0.202) + 36.0(0.446)] (t − 20°C)

t = 560°C

Chapter 5

1. a) 674 mm Hg; 563 mm Hg; 439 mm Hg; 512 mm Hg

   b) 5.42×10^{-3}; 0.152 g   c) 1.13 g/liter   d) 50.5

2. 22.8 liters O_2, 13.0 liters NO_2 at same T, P

3. 648 mm Hg

4. 0.707

5. 515 m/sec

6. 0.786; 0.214

7. 6.70 atm

8. a) 0.490   b) 0.490; 14.7 g   c) 12.0 liters   d) 4.22 liters

9. a) 63.4   b) 2.83 g/liter

10. a) 1.59 atm; 0.64 atm   b) 2.23 atm

11. 11.2 sec

12. a) $n_t = \dfrac{PV}{RT} = 0.0350$; n $PH_3 = \dfrac{0.680}{34.0} = 0.0200$

   1 mole $PH_3 \rightarrow 1.75$ moles products; 4 moles $PH_3 \rightarrow 7$ moles products

   b) $4 PH_3(g) \rightarrow P_4(g) + 6H_2(g)$

Chapter 6

1. 6.41×10^{-12} erg; 82800 Å

2. $1s^2 2s^2 2p^5$; $1s^2 2s^2 2p^6 3s^2 3p^3$; $1s^2 2s^2 2p^6 3s^2 3p^6 4s^2 3d^7$

3.

	1s	2s	2p	3s	3p	4s	3d
F	(↑↓)	(↑↓)	(↑↓) (↑↓) (↑)				
P	(↑↓)	(↑↓)	(↑↓) (↑↓) (↑↓)	(↑↓)	(↑) (↑) (↑)		
Co	(↑↓)	(↑↓)	(↑↓) (↑↓) (↑↓)	(↑↓)	(↑↓) ↑↓) ↑↓)	(↑↓)	(↑↓) (↑↓) (↑) (↑) (↑)

4. F 1,0,0,+1/2 } 1s   2,1,+1,+1/2
       1,0,0,−1/2 }   2,1,+1,−1/2
                             2,1, 0,+1/2 } 2p
       2,0,0,+1/2 } 2s   2,1, 0,−1/2
       2,0,0,−1/2 }   2,1,−1,+1/2

    P 3,1,+1,+1/2
       3,1, 0,+1/2
       3,1,−1,+1/2

5. a) ~1.8 g/cc   b) Ca; Ba; Sr   c) $Ga_2(SeO_4)_3$

6. a) 8   b) $1s^2 2s^2 2p^4$

c) 1s   2s   2p
   (↑↓) (↑↓) (↑↓) (↑) (↑)

d) $\left.\begin{array}{l} 1,0,0,+1/2 \\ 1,0,0,-1/2 \end{array}\right\}$ 1s $\left.\begin{array}{l} 2,1,+1,+1/2 \\ 2,1,+1,-1/2 \\ 2,1, 0,+1/2 \\ 2,1,-1,+1/2 \end{array}\right)$ 2p

$\left.\begin{array}{l} 2,0,0,+1/2 \\ 2,0,0,-1/2 \end{array}\right\}$ 2s

7. a) ground   b) excited   c) excited   d) impossible   e) impossible

8. b, c

9. a) $[Ar]\, 4s^2 3d^5$

b)   4s   3d
   (↑↓) (↑)(↑)(↑)(↑)(↑); 5 unpaired electrons

c) smaller than Sc, La, Cs; larger than Cl

d) Sc, La, Cs; Cl

e) $CsTcS_4$

10. a) 1.634×10^{-11} erg/atom   b) 235.1 kcal/mole   c) 1217 Å

11. $E = -\dfrac{2.179 \times 10^{-11} \text{ erg}}{n^2}$; $(-2.179, -0.545, -0.242, -0.136, -0.087, -0.061)\, 10^{-11}$ erg

$\Delta E = \dfrac{6.626 \times 10^{-27} \text{ erg sec} \times 3.00 \times 10^{10} \text{ cm/sec}}{1.026 \times 10^{-5} \text{ cm}} = 1.937 \times 10^{-11}$ erg

n = 3 to n = 1; UV

12. $\Delta E = 313.6\, \dfrac{\text{kcal}}{\text{mole}} \times \dfrac{6.949 \text{ erg/atom}}{1 \text{ kcal/mole}} = 2.179 \times 10^{-11}$ erg/atom

$= \dfrac{6.627 \times 10^{-27} \text{ erg sec} \times 3.00 \times 10^{10} \text{ cm/sec}}{2.179 \times 10^{-11} \text{ erg/atom}} = 9.12 \times 10^{-6}$ cm = 912 Å

Chapter 7

1. a) LiBr   b) Li_2O c) Li_2SO_4 d) SrS   e) SrI_2 f) $Sr(NO_3)_2$

g) $AlPO_4$ h) Al_2O_3 i) $Al(NO_3)_3$

2. a) :F̈ - C̈l: b) :B̈r - Ö: c) :C̈l - Ö - C̈l: d) :Ö - N̈ = Ö

e)
:Ö   Ö:
   \\   /
     Se
     ‖
    :O:

3. a) linear   b) linear   c) bent, 109° angle   d) bent, 120° angle

e) equilateral triangle with Se at center

4. a, c

5. sp^2, sp^3

6. a) $1s^2$ b) $1s^2 2s^2 2p^6$ c) $1s^2 2s^2 2p^6 3s^2 3p^6 3d^5$

7. a) Be^{2+} b) Co^{2+} c) F

8. $O-O$, $N-O$, $P-O$, $Ge-O$

9. d, e

10.

σ_{2s}^{b}	σ_{2s}^{*}	π_{2p}^{b}	π_{2p}^{b}	σ_{2p}^{b}	π_{2p}^{*}	π_{2p}^{*}	σ_{2p}^{*}

a) (↑↓) (↑↓) (↑↓) (↑↓) (↑↓) (↑) (↑) ( )

b) (↑↓) (↑↓) (↑↓) (↑↓) (↑↓) (↑↓) (↑) ( )

11. $:\ddot{O}-H$; 180°; dipole $:\ddot{O}-\ddot{Cl}-\ddot{O}:$; 109°, dipole

$\quad\quad\quad\quad\quad\quad\quad\quad\quad\quad\quad\quad\quad\quad :\ddot{O}:$

H
|
$H-C-\ddot{O}-H$; 109°; dipole $:\ddot{O}\quad\quad\ddot{O}:$
|
H

$\quad\quad\quad\quad\quad\quad\quad\quad\quad\quad\quad\quad\quad\quad\quad S\quad$; 120°; nonpolar

$:\ddot{S}-C\equiv N:$; 180°; dipole $:\underset{\|}{O}:$

12. a)

H   H
   \   /
    C   Ö
   /   \   /   \
  H   C   H
          ||
         :O:

          109°: $H-C-H$, $H-C-C$, $C-O-H$

          120°: $C-C-O$, $O-C-O$

b) sp^3, sp^2

13. a) $NaNO_3$, $1s^2 2s^2 2p^6$, $:O\quad\quad O:$ b) K_2CO_3, [Ar], $:O\quad\quad O:$

$\quad\quad\quad\quad\quad\quad\quad\quad\quad\quad\quad\quad\quad\quad N\quad\quad\quad\quad\quad\quad\quad\quad\quad\quad\quad\quad\quad C$
$\quad\quad\quad\quad\quad\quad\quad\quad\quad\quad\quad\quad\quad\quad \|\quad\quad\quad\quad\quad\quad\quad\quad\quad\quad\quad\quad\quad \|$
$\quad\quad\quad\quad\quad\quad\quad\quad\quad\quad\quad\quad\quad\quad :O:\quad\quad\quad\quad\quad\quad\quad\quad\quad\quad\quad\quad :O:$

$\quad\quad\quad\quad\quad\quad\quad\quad\quad\quad\quad\quad\quad\quad :\ddot{O}:\quad\quad\quad\quad\quad\quad\quad\quad\quad\quad\quad :\ddot{O}:$
$\quad\quad\quad\quad\quad\quad\quad\quad\quad\quad\quad\quad\quad\quad |\quad\quad\quad\quad\quad\quad\quad\quad\quad\quad\quad\quad |$
c) $Ca_3(PO_4)_2$, [Ar], $:\ddot{O}-P-\ddot{O}:$ d) $NiSO_4$, [Ar] $3d^8$, $:\ddot{O}-S-\ddot{O}:$
$\quad\quad\quad\quad\quad\quad\quad\quad\quad\quad\quad\quad\quad\quad |\quad\quad\quad\quad\quad\quad\quad\quad\quad\quad\quad\quad\quad |$
$\quad\quad\quad\quad\quad\quad\quad\quad\quad\quad\quad\quad\quad\quad :O:\quad\quad\quad\quad\quad\quad\quad\quad\quad\quad\quad\quad :O:$

Chapter 8

1. CaF_2; ionic solid, conductor in liquid state

Br_2, NO_2; molecular

SiO_2; macromolecular solid

Al; metallic solid, conductor in both liquid and solid states

2. Dispersion forces in each case; dipole forces in NO; H-bonds in HF and NH_3

3. a) ICl   b) CCl_4 c) H_2O d) Ar

4.

$$H_3C\diagdown \atop H_3C\diagup C=C \diagup ^H _{\diagdown H} \quad , \quad H\diagdown \atop H_3C\diagup C=C \diagup ^{CH_3} _{\diagdown H} \quad , \quad H_3C\diagdown \atop H\diagup C=C \diagup ^{CH_3} _{\diagdown H} \quad , \quad H\diagdown \atop H\diagup C=C \diagup ^{CH_2CH_3} _{\diagdown H}$$

5. a) paraffin   b) aromatic   c) olefin   d) acetylene   e) olefin

6. a) alcohol   b) ester   c) aldehyde   d) acid   e) ether   f) ester

7. a)

$$\cdot \overset{\displaystyle H}{\underset{\displaystyle H}{C}} - \overset{\displaystyle CN}{\underset{\displaystyle H}{C}} - \overset{\displaystyle H}{\underset{\displaystyle H}{C}} - \overset{\displaystyle CN}{\underset{\displaystyle H}{C}} \cdot$$

b) $-O-CH_2-CH_2-O-\overset{O}{\overset{\|}{C}}-CH_2-\overset{O}{\overset{\|}{C}}-O-CH_2-CH_2-O-$

8. H bonding: b, c

highest boiling: e

b) is an alcohol,   c) an acid,   f) an ether

isomers: a, c, d, e, f

9. a) KI ionic (higher bpt), I_2 molecular

b) C macromolecular (higher bpt), O_2 molecular

c) CO_2 molecular, CS_2 molecular (higher bpt)

d) Hg metallic (higher bpt), He molecular

10. a, c, f

11. a) paraffin   b) olefin   c) acetylene   d) aromatic

12. a)

$$H_3C-\overset{H}{\overset{|}{C}}=O \quad \text{(aldehyde)}$$

b) $H_3C-\overset{O}{\overset{\|}{C}}-OH$ (acid), $H-\overset{O}{\overset{\|}{C}}-O-CH_3$ (ester)

c) $H_3C-O-CH_3$ (ether), H_3C-CH_2-OH (alcohol)

13. a) $H-\overset{}{\underset{Cl}{\overset{|}{C}}}=\overset{}{\underset{Cl}{\overset{|}{C}}}-H$

b) $HO-\overset{O}{\overset{\|}{C}}-CH_2-CH_2-\overset{O}{\overset{\|}{C}}-OH$ and $HO-CH_2-CH_2-OH$

Chapter 9

1. 0.327 kcal

2. a) 260 mm Hg   b) 93 mm Hg

3. 170 mm Hg

4. 1.25 Å

5. 1, solid; 2, solid + liquid; 3, liquid; 4, solid, liquid, gas; 5, gas; less

6. a) log vp   1.255   1.740   2.173

      1/T   0.003413   0.003195   0.003003

   d) -2240   e) -10,300 cal/mole

7. 3.26°, 6.53°, 9.82°

8. 460°K

9. a) 0.160 kcal   b) 0.200 kcal   c) 1.08 kcal

10. a) 183 mm Hg   b) 75 mm Hg

11. a) 140 mm Hg   b) 40°C   c) 118°C

12. a) 24.1 (Å)^3; 2.41 X 10^{-23} cm^3 b) 2   c) 17.3 X 10^{-23} g   d) 7.18 g/cm^3

13.

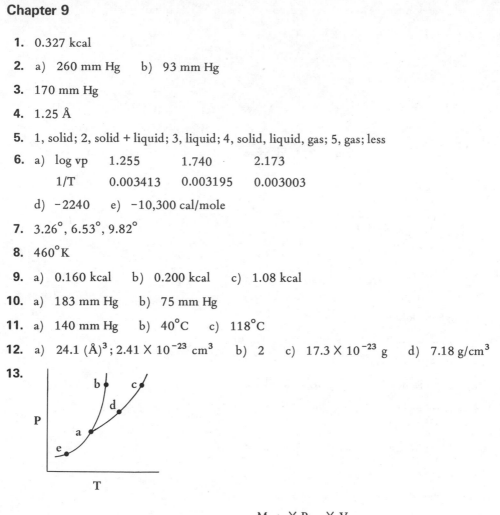

14. $g = $ wt vapor + wt air final - wt air orig $= \dfrac{M_{calc} \times P_{bar} \times V}{R(373)}$

$$= \dfrac{M_{true} \times P_{bar} \times V}{R(373)} + \dfrac{29.0(P \text{ air final})V}{R(298)} - \dfrac{29.0(P \text{ air orig})V}{R(298)}$$

$$\therefore \dfrac{(M_{calc} - M_{true})P_{bar}}{373} = \dfrac{29.0(P \text{ air final} - P \text{ air orig})}{298}$$

$M_{calc} - M_{true} = 29.0(P \text{ air final} - P \text{ air orig}) \times 373/298$

but, P air final + v.p. = P air orig

$\therefore M_{calc} - M_{true} = -29.0 \dfrac{(v.p.)}{P_{bar}} \times \dfrac{373}{298}$

Chapter 10

1. a) 37.5, 62.5   b) 0.252, 0.748   c) 18.8

2. 0.201

3. 0.0400

4. 0.050 mole/liter

5. a) CH_3OH b) Ar   c) B

6. a) 15 mm Hg   b) 19 mm Hg   c) 34 mm Hg

7. a) $-2.48°C$, $100.69°C$ b) 255

8. 0.995 atm

9. a) 0.278   b) 1020 g   c) 970 g   d) 4.90, 95.1   e) 0.286
   f) 5.13×10^{-3}

10. $-0.532°C$, $100.15°C$, 6.69 atm, 0.122 mm Hg

11. $---$, 0.222, $---$, 0.135
    142, $---$, 0.374, $---$
    55.5, 0.500, $---$, $---$
    $---$, 0.135, 0.0338, $---$

12. a) 2.00 liters   b) 3.00 liters

13. a) O_2 in both   b) CCl_4; NaCl   c) 2-methylnaphthalene in both
    d) C_6H_{14}; H_2O_2

14. a) 359   b) 193   c) 1530

Chapter 11

1. a) 2   b) 3   c) 5   d) 3   e) 3

2. a) $-0.0372°C$ b) $-0.0558°C$ c) $-0.0930°C$ d) $-0.0558°C$
   e) $-0.0558°C$

3. a) 800 g   b) 80.0

4. a) 1.2×10^{-4}, 0   b) 0.8×10^{-4}, 0.4×10^{-4}

5. 7.5 atm

6. a) 1.2   b) 1.0, 2.0, 1.5   c) 20%

7. a) 65   b) 0.065   c) 1.1×10^{-3} M

8. a) 2.2×10^{-3} mole Na_2CO_3 **b)** 0.80×10^{-3} mole $Ca(OH)_2$,

0.3×10^{-3} mole Na_2CO_3

c) none **d)** none

9. no. g O_2 in one liter water $= (1.2 \times 10^{-3})(32.0) = 0.038$

no. liters water required $= \dfrac{0.140 \text{ g}}{0.038 \text{ g/liter}} = 3.7$ liters

dilution factor $= 4.7/1.0$; i.e., volume is 4.7 times that of waste

Chapter 12

1. a) -26.3 kcal **b)** -56.6 kcal

2. a) $-$ **b)** $-$ **c)** $+$

3. 2.39 cal/$°$K; 22.2 cal/$°$K

4. a) -0.0403 kcal/$°$K **b)** $+7.2$ kcal **c)** $521°$K

5. a) -22.0 kcal **b)** -8.0 kcal **c)** -0.0470 kcal/$°$K **d)** $681°$K

6. a) -33.2 kcal **b)** -0.0937 kcal/$°$K **c)** -5.3 kcal

7. $T = \Delta H / \Delta S = 3020°$K, $840°$K, $700°$K; 3rd reaction is most feasible

8. Suppose CO were formed; the further reaction: $CO(g) + \dfrac{1}{2}O_2(g) \rightarrow CO_2(g)$ would occur since $\Delta G^{1 \text{ atm}} = -61.5$ kcal at $25°$C. One can show that as T increases, ΔG becomes less negative, and reaction is less likely to occur.

9. Consider: $CrCl_3(s) \rightarrow CrCl_2(s) + \dfrac{1}{2}Cl_2(g)$

$\Delta H = \Delta H_f CrCl_2 - \Delta H_f CrCl_3 = -94.6 \text{ kcal} + 134.6 \text{ kcal} = +40.0 \text{ kcal}$

$\Delta G = \Delta G_f CrCl_2 - \Delta G_f CrCl_3 = +32.8 \text{ kcal at } 25°C, 1 \text{ atm}$

$\Delta S = 7.2 \text{ kcal}/298°K = 0.024 \text{ kcal}/°K$

$\Delta G_T = 40.0 \text{ kcal} - 0.024 \text{ T}$; reaction reverses at about $1700°$K

For all such reactions, both ΔH and ΔS are positive, as in the above example.

Chapter 13

1. $\dfrac{[PCl_3]^4}{[P_4] \times [Cl_2]^6}$; $\dfrac{[CO_2]^3}{[CO]^3}$

2. $1.36; 0.400$

3. a) 3.0×10^{-2} **b)** 9.5×10^{-5}

4. a) $\leftarrow$ **b)** 1.7 mole/liter **c)** 1.0 mole/liter **d)** $0.66, 1.37$

e) $0.090, 0.025$

5. 170

6. a) $\rightarrow$ b) $\rightarrow$ c) $\rightarrow$ d) $\rightarrow$ e) $\leftarrow$ f) 0 g) $\rightarrow$

7. 5.1×10^9

8. a) $\dfrac{[AB]^2}{[A_2] \times [B_2]}$ b) 1.67 mole/liter c) 1.18 mole/liter

9. a) $\leftarrow$ b) 0.84 c) $\rightarrow$, 0.16 d) $\rightarrow$, 1.07

10. a) 0.39 b) 2.6 c) 1.6 d) 0.0034

11. 1.50 mole/liter

12. a) 0.028 b) HBr more stable (ΔG_f more negative at $1000°K$)

Chapter 14

1. 1st, 2nd, zero

2. a) 0.052 b) 140 sec

3. 32 sec

4. a) 2.5×10^{-2}/min b) 8.6 kcal c) $41°C$

5. b

6. 5.8 kcal

7. 2

8. a) 11 kcal b) 0.188/min c) $60°C$ d) 150 min, 46 min e) 0.38
   f) 73 min

9. a) 1, 2 b) $1.25 \text{ lit}^2 \text{mole}^{-2} \text{min}^{-1}$ c) 1.25 mole/lit min d) 2nd

Chapter 15

1. 20.95, 23.1, 0.2095 atm, 0.00856 mole/liter

2. 2600 Å

3. 47.3%

4. a) 53.6% b) 0.0160 c) 6.48×10^{-4} d) 1.60%, 0.993%, 1.60×10^4,
   1.60×10^7

5. a) 4.96×10^{-12} b) yes, no

Chapter 16

1. a) $Ag^+(aq) + Cl^-(aq) \rightarrow AgCl(s)$

   b) $Cu^{2+}(aq) + 2OH^-(aq) \rightarrow Cu(OH)_2(s)$

   d) $Cu^{2+}(aq) + 2OH^-(aq) \rightarrow Cu(OH)_2(s); Ba^{2+}(aq) + SO_4{}^{2-}(aq) \rightarrow BaSO_4(s)$

2. a) 0.060, 0.12, 0.090, 0.090   b) Cl^- c) 0.045, 0.090   d) 0.015, 0.12,

   0.090, ~ 0

3. a) $[Cu^{2+}] \times [CO_3{}^{2-}]$ b) $[Ag^+]^2 \times [CrO_4{}^{2-}]$ c) $[Al^{3+}] \times [OH^-]^3$

   d) $[Zn^{2+}]^3 \times [PO_4{}^{3-}]^2$

4. 1.4×10^{-5}, 1.4×10^{-4}, 9×10^{-22}, 3×10^{-8}

5. a) 5×10^{-12} b) yes   c) $Al(OH)_3$

6. a) 46.7   b) 36.0

7. add Cl^-, then $SO_4{}^{2-}$, then OH^-

8. a) $BaCl_2$ b) $Ba(NO_3)_2$ c) $NaOH$

9. a) $Al^{3+}(aq) + 3OH^-(aq) \rightarrow Al(OH)_3(s)$

   b) 0.040   −0.020   0.020   0.040

    0.12   0.12   0.24

    0.060   0.060   0.12

    0.060   −0.060   0.000   0.000

10. a) 7×10^{-5}, 7×10^{-3} b) no   c) 6×10^{-7} d) 1%

11. a) 60.4   b) 0.751 g

Chapter 17

1. _____, 1.0×10^{-10}, 4.00

   5.0×10^{-9}, _____, 8.30

   5.0×10^{-4}, 2.0×10^{-11}, ____

2. a) 9.0×10^{-7} b) 2.5×10^{-5}

3. a) 2.6×10^{-2}, 2.6%   b) 8.0×10^{-3}, 8.0%

4. 1.4×10^{-11}

5. 1.2×10^{-6}

6. a) SB   b) SA   c) WA   d) WB   e) WA

7. acids: Al^{3+}, Fe^{3+} bases: $CO_3{}^{2-}$, $PO_4{}^{3-}$ neutral: Na^+, NO_3^-

8. a) N   b) A   c) B   d) B   e) A

9. a) $HCl(aq) \rightarrow H^+(aq) + Cl^-(aq)$ b) $CO_3{}^{2-}(aq) + H_2O \rightarrow HCO_3^-(aq) + OH^-(aq)$

    c) $NH_4{}^+(aq) \rightarrow NH_3(aq) + H^+(aq)$ d) $NH_3(aq) + H_2O \rightarrow NH_4{}^+(aq) + OH^-(aq)$

10. $H_2SO_4 > H_2SeO_4 > H_2SeO_3$

11. a) BA: HCO_3^-, H_2O, HCO_3^- BB: OH^-, HCO_3^-, H_2O

    b) LA: Ag^+, BeF_2 LB: NH_3, F^-

12. a) 4.0×10^{-8} b) 1.4×10^{-5} c) 2.8×10^{-3} d) $4.85, 7.1 \times 10^{-10}$

    e) 2.5×10^{-5} f) 1.6×10^{-3} g) 6.2×10^{-12}, 11.21

13. a) SA: HBr   SB: KOH   WA: HNO_2, $Cu(H_2O)_4{}^{2+}$ WB: NO_2^-, $CO_3{}^{2-}$ N: Br^-, K^+

    b) $HNO_2(aq) \rightarrow H^+(aq) + NO_2^-(aq)$; $Cu(H_2O)_4{}^{2+} \rightarrow H^+(aq) + Cu(H_2O)_3(OH)^+(aq)$

    c) $NO_2^-(aq) + H_2O \rightarrow HNO_2(aq) + OH^-(aq)$; $CO_3{}^{2-}(aq) + H_2O \rightarrow HCO_3^-(aq) + OH^-(aq)$

    d) BA: H_2O BB: NO_2^-, $CO_3{}^{2-}$

    e) KBr, KNO_2 or K_2CO_3, $CuBr_2$

    f) HNO_3; HNO or HPO_2

Chapter 18

1. a) 4.5×10^{10} b) 1.6   c) 2×10^8

2. a) 0.200   b) 208 g

3. a) 2.5   b) 2.5×10^{-6} c) 5.60   d) $1.8, 1.8 \times 10^{-6}$, 5.74

4. 62.0 g, 1.0; 31.0 g, 0.80

5. a) $H^+(aq) + OH^-(aq) \rightarrow H_2O$; $2H^+(aq) + Sr(OH)_2(s) \rightarrow Sr^{2+}(aq) + 2H_2O$

    b) $HF(aq) + OH^-(aq) \rightarrow F^-(aq) + H_2O$

    c) $H^+(aq) + NH_3(aq) \rightarrow NH_4{}^+(aq)$

    $2H^+(aq) + CaCO_3(s) \rightarrow Ca^{2+}(aq) + CO_2(g) + H_2O$

6. a) 1×10^{-5} b) 1×10^{-9} c) any   d) none

7. a) add acid, test for $CO_2(g)$ b) heat with base, test for $NH_3(g)$

    c) heat with acid, test for $H_2S(g)$

8. a) neutralize with HNO_3, evaporate

    b) ppt. $Mg(OH)_2$ with NaOH, neutralize with HNO_3, evaporate

9. a) $H^+(aq) + C_2H_3O_2^-(aq) \rightarrow HC_2H_3O_2(aq)$ b) 5.6×10^4 c) no   d) 40.0

    e) 59.0 g

10. a) 1.0×10^{-4} b) 0.50   c) 0.50   d) 0.33

11. a) · $MS(s) + 2H^+(aq) \rightarrow M^{2+}(aq) + H_2S(g)$

    b) use Law of Multiple Equilibria, add three equations

    c) test for H_2S after adding H^+; MSO_4 would not react

    d) heat with HCl or HNO_3, evaporate

Chapter 19

1. $Cu(NH_3)_4{}^{2+}$; $Cr(NH_3)_5Cl^{2+}$; $4NH_3$, $2Cl^-$; Co^{3+}, $4NH_3$, $2Cl^-$

2. a) 6   b) octahedral   c) last two complexes have geometrical isomers

3. a) two 3d, one 4s, three 4p; one 4s, three 4p, two 4d

    d)   3d   4s   4p   4d

    (↑↓) (↑↓) (↑↓) (↑↓) (↑↓) (↑↓) (↑↓) (↑↓) (↑↓) (↑)( )( )( )( )

    (↑↓) (↑↓) (↑) (↑) (↑) (↑↓) (↑↓) (↑↓) (↑↓) (↑↓) (↑↓) ( )( )( )

4. a) 3×10^{-6} b) 3×10^{-14}

5. a) $Zn^{2+}(aq) + 4OH^-(aq) \rightarrow Zn(OH)_4{}^{2-}(aq)$

    b) $Zn(OH)_2(s) + 2OH^-(aq) \rightarrow Zn(OH)_4{}^{2-}(aq)$

6.   (↑)(↑)   ( )( )

    (↑↓) (↑) (↑)   (↑↓) (↑↓) (↑↓)

7. a) $Mn(NH_3)_3Br_3{}^-$ b) $K[Mn(NH_3)_3Br_3]$ c) KCl   d) octahedral, 2 isomers

    e)   3d   4s   4p

    (↑↓) (↑↓) (↑) (↑↓) (↑↓)   (↑↓)   (↑↓) (↑↓) (↑↓)

8. a) $Fe^{3+}(aq) + 6CN^-(aq) \rightarrow Fe(CN)_6{}^{3-}(aq)$

    b) $Fe(OH)_3(s) + 6CN^-(aq) \rightarrow Fe(CN)_6{}^{3-}(aq) + 3OH^-(aq)$

    c) 1×10^{31}; 5×10^{-7} d) no

Chapter 20

1. a) +1   b) +5, -2   c) +1, +3, -2   d) +6, -2   e) +2, -2

2. $2NO_3^-(aq) + 6I^-(aq) + 8H^+(aq) \rightarrow 2NO(g) + 3I_2(s) + 4H_2O$

    $2NO_3^-(aq) + 6I^-(aq) + 4H_2O \rightarrow 2NO(g) + 3I_2(s) + 8OH^-(aq)$

3. a) 3.28   b) 69.4   c) 36.8

4. a) 0.126   b) 11400 sec

5. a) Ni anode, Pt cathode. Electrons move out of Ni to Pt. Cations move to Pt, anions to Ni.

    b) Zn cathode, Pt anode. Electrons move out of Pt to Zn. Cations move to Zn, anions to Pt.

6. 36.0 g, 17.3 g

7. a) $5Fe^{3+}(aq) + Mn^{2+}(aq) + 4H_2O \rightarrow 5Fe^{2+}(aq) + MnO_4^-(aq) + 8H^+(aq)$

    b) Fe: +3 to +2; Mn: +2 to +7; O: −2; H: +1

    c) Pt cathode surrounded by Fe^{3+}, Fe^{2+}; Pt anode surrounded by Mn^{2+}, MnO_4^-

    d) 4.98   e) 19.3   f) 2090 sec

Chapter 21

1. a) $Ba^{2+} < Sn^{4+} < Fe^{3+} < Co^{3+}$ b) 1.67 V   c) −77.0 kcal   d) 4×10^{56}

    e) V, E, V

2. a) +0.43 V   b) 2.6×10^{-3}

3. 0.655 M

4. a) $Ni^{2+}(aq) + H_2(g) \rightarrow Ni(s) + 2H^+(aq)$

        $Cu^{2+}(aq) + H_2(g) \rightarrow Cu(s) + 2H^+(aq)$

        $Ni^{2+}(aq) + 2Ag(s) \rightarrow Ni(s) + 2Ag^+(aq)$

        $Cu^{2+}(aq) + 2Ag(s) \rightarrow Cu(s) + 2Ag^+(aq)$

    b) 2nd

    c) +11.5 kcal, +48.4 kcal, +21.2 kcal; 3×10^{-9}, 3×10^{-36}, 3×10^{-16}

    d) Ni, H_2, Cu, Ag; Cu^{2+}, Ag^+

5. a) −0.02 V   b) 5.4 M

Chapter 22

1. a) $^{226}_{88}Ra$ b) $^{214}_{83}Bi$ c) $^{1}_{0}n$

2. a) $0.0693 \ hr^{-1}$ b) 0.76 g   c) 14.7 hr

3. a) 0.15876   b) 148   c) 7.78

4. a) -6.0×10^{-3} b) -6.0×10^{-3} c) -2.7×10^{-5} d) −5.6

    e) -5.8×10^5 kcal

5. a) E   b) P   c) E

6. a) $^{210}_{84}Po \rightarrow ^{4}_{2}He + ^{206}_{82}Pb$ b) -5.8×10^{-3} amu   c) -1.2×10^8 kcal

    d) 4.95×10^{-3} e) 465 days

7. $^{7}_{3}Li$

8. about 1600 yrs

Chapter 23

1. b

2.

a) $H_2N-\overset{\overset{\displaystyle H}{|}}{\underset{\underset{\displaystyle R}{|}}{C}}-\overset{\overset{\displaystyle O}{||}}{C}-\overset{\overset{\displaystyle H}{|}}{N}-\overset{\overset{\displaystyle H}{|}}{\underset{\underset{\displaystyle CH_3}{|}}{C}}-\overset{\overset{\displaystyle O}{||}}{C}-OH,$ where $R = CH_3-\overset{\overset{\displaystyle H}{|}}{\underset{\underset{\displaystyle CH_3}{|}}{C}}-$

b) $H_2N-\overset{\overset{\displaystyle R}{|}}{\underset{\underset{\displaystyle H}{|}}{C}}-\overset{\overset{\displaystyle O}{||}}{C}-\overset{\overset{\displaystyle H}{|}}{N}-\overset{\overset{\displaystyle R'}{|}}{\underset{\underset{\displaystyle H}{|}}{C}}-\overset{\overset{\displaystyle O}{||}}{C}-\overset{\overset{\displaystyle H}{|}}{N}-CH_2-\overset{\overset{\displaystyle O}{||}}{C}-OH$

where $R = H_2N-(CH_2)_3-CH_2-$, and $R' = HS-CH_2-$

3. Ser − Ala − Cys − Pro − Ser − Ala − Cys

4.

a) $H-\overset{\overset{\displaystyle Cl}{|}}{\underset{\underset{\displaystyle OH}{|}}{C}}-\overset{\overset{\displaystyle H}{|}}{\underset{\underset{\displaystyle OH}{|}}{C}}-H$ b) $H_3C-CHCl-CH_2-CH_3$ c) $H-\overset{\overset{\displaystyle H}{|}}{\underset{\underset{\displaystyle Cl}{|}}{C}}-\overset{\overset{\displaystyle OH}{|}}{\underset{\underset{\displaystyle H}{|}}{C}}-CH_3$

5.

$H_2N-CH_2-\overset{\overset{\displaystyle O}{||}}{C}-\overset{\overset{\displaystyle H}{|}}{N}-CH_2-\overset{\overset{\displaystyle O}{||}}{C}-\overset{\overset{\displaystyle H}{|}}{N}-\overset{\overset{\displaystyle R}{|}}{\underset{\underset{\displaystyle H}{|}}{C}}-\overset{\overset{\displaystyle O}{||}}{C}-OH$

$H_2N-CH_2-\overset{\overset{\displaystyle O}{||}}{C}-\overset{\overset{\displaystyle H}{|}}{N}-\overset{\overset{\displaystyle R}{|}}{\underset{\underset{\displaystyle H}{|}}{C}}-\overset{\overset{\displaystyle O}{||}}{C}-\overset{\overset{\displaystyle H}{|}}{N}-CH_2-\overset{\overset{\displaystyle O}{||}}{C}-OH$ where R is $H-\overset{\overset{\displaystyle OH}{|}}{\underset{\underset{\displaystyle H}{|}}{C}}-$

$H_2N-\overset{\overset{\displaystyle R}{|}}{\underset{\underset{\displaystyle H}{|}}{C}}-\overset{\overset{\displaystyle O}{||}}{C}-\overset{\overset{\displaystyle H}{|}}{N}-CH_2-\overset{\overset{\displaystyle O}{||}}{C}-\overset{\overset{\displaystyle H}{|}}{N}-CH_2-\overset{\overset{\displaystyle O}{||}}{C}-OH$